# Experimental Soil Biology and Fertility

Soil fertility and plant nutrition is an applied science that integrates knowledge across all disciplines of soil and plant sciences to provide nutrients effectively and efficiently to plants. Efficient use of nutrients is required not only to maximize agricultural production but also to protect air, soil, and water quality as well as the natural resources involved in providing fertilizers to support agricultural production.

This book, *Experimental Soil Biology and Fertility*, by Dr. A. S. Mailappa, is truly a tour de force of condensation of the essentials of scientific knowledge and approaches to soil science and discusses the various aspects of soil fertility and crop nutrition with a focus on collection, preparation, and analysis of essential plant nutrients in soil, plant, and water.

This book is unique, written in a simple and lucid manner and covering all aspects of soil fertility and biology in comprehensive chapters. This book is organized to facilitate rapid location of information, while being written in a readable style. The topics and discussion in this self-contained book are practical and user-friendly, yet comprehensive enough to cover material presented in upper-level soil and plant science courses. It allows practitioners with general background knowledge to feel confident applying the principles presented to soil/crop production systems.

**Readership**: students / teachers / researchers / practitioners of agricultural universities / institutes, engaged in teaching, research and extension activities related to agriculture, horticulture, forestry, and other allied disciplines.

# Experimental Soil Biology and Fertility

A.S. Mailappa

CRC Press
Taylor & Francis Group
Boca Raton London New York

CRC Press is an imprint of the
Taylor & Francis Group, an **informa** business

First edition published 2024
by CRC Press
6000 Broken Sound Parkway NW, Suite 300, Boca Raton, FL 33487-2742

and by CRC Press
4 Park Square, Milton Park, Abingdon, Oxon, OX14 4RN

*CRC Press is an imprint of Taylor & Francis Group, LLC*



ISBN: 9781032553207 (hbk)
ISBN: 9781032553214 (pbk)
ISBN: 9781003430100 (ebk)

DOI: 10.1201/9781003430100

Typeset in Times
by Newgen Publishing UK

# Contents

# Foreword

Knowledge gained from scientific research over the years and its successful applications have made a tremendous impact on agricultural production. The introduction of high-yielding varieties has resulted in greater demands on plant nutrients which cannot be met from the inherent soil fertility. The natural resource conservation and its management through the minimum external input has become a challenge to the scientific community.

Rational use of these inputs is imperative, particularly in developing countries where the financial limitations of the farmers are a major constraint. Even in developed countries, one cannot overlook the rational and economic use of such inputs as fertilizers because of financial considerations as well as the ever-growing concern for environmental pollution.

Soil fertility and plant nutrition encompasses the management of essential elements necessary for plant growth, typically to achieve selected management objectives. Soil fertility plays a vital role in natural systems, especially plant production for human uses. A crucial aspect of plant nutrition is therefore to preserve the fertility of soils so that they can deliver nutrients at the right time and in the right quantity for growing plants.

Soil testing and plant analysis in conjunction with fertilizer trials in the field are indispensable tools in respective research and for the formulation of fertilizer recommendations. The need for improvement and coordination of these services is of great importance, particularly in developing countries where resources must be put to maximum use.

Experimental methods for evaluating nutrient status in order to obtain better plant growth and increased yields are constantly being developed and improved. New analytical techniques and procedures for soil and plant analysis have been invented and tested in many countries and laboratories. There have been innovations in data processing leading to the preparation of more refined and specific fertilizer recommendations.

Analytical and instrumental methods are the foundation of a scientific discipline. They play an important role in the analysis of soil, plants, and water. Advanced analytical techniques and experimental methods are required that characterize the fitness or quality of soils to perform various functions, such as providing a medium for plant growth, recycling waste products, and regulating and storing water, energy, and nutrients.

This book, *Experimental Soil Biology and Fertility*, by Dr. A.S. Mailappa, Associate Professor (Soil Science and Agricultural Chemistry), College of Horticulture & Forestry, Central Agricultural University (Imphal), Pasighat, Arunachal Pradesh, is an authentic treatise, presented in a simple and lucid manner that the reader can easily understand various basic aspects of soil biology and fertility.

The book presents the basic concepts and approaches to outline the application of such techniques used in experimental soil biology and fertility to address a wide range of disciplines, standard methodologies, and research techniques, with an emphasis on modern instrumental methods. Emphasis is placed on sampling and sample preparation, followed by a description of the possible instrumental methods available for the analysis of physico-chemical and biological properties, in addition to soil available nutrients and plant nutrient contents.

This book has been written in such a way that the reader shall easily understand various aspects of analysis of test samples of soil, water, and plants. At the end of each of the 41 well-defined chapters, the author has also put some questions (things to learn) to assist readers' in-depth understanding of the subject, in addition to inspiring them to expand their ideas and horizons in all possible ways.

The author needs to be complimented for his efforts in bringing out this important book. I am confident that the book will be very useful for undergraduate and postgraduate students, researchers of soil science and other disciplines related to agriculture, horticulture, and forestry, in particular, and the scientific community, as a whole. In pursuit of maintenance of soil quality, this publication will add value and deserves to be the standard textbook on experimental soil biology and fertility in agricultural/horticultural/forestry colleges and universities across the world.

**(Anupam Mishra)**

**Dr. Anupam Mishra**
*Vice Chancellor*
*Central Agricultural University*
*Imphal*

# Preface

The diverse challenges and constraints of population growth, increasing food, feed, and fodder needs, natural resource degradation, climate change, new parasites, slow growth in farm income, and new global trade regulations demand a paradigm shift in formulating and implementing agricultural research programs.

In the recent years, a declining trend of total factor productivity and compound growth rates of major crops and low nutrient use efficiency have been observed primarily due to deterioration of soil health. The main reasons for soil health deterioration are the wide nutrient gap between nutrient demand and supply, high nutrient turnover in the soil–plant system coupled with low and imbalanced fertilizer use, decline in organic matter status, emerging deficiencies of secondary and micronutrients, nutrient leaching and fixation problems, impeded drainage, soil pollution, soil acidity, salinization and sodification, etc.

The widespread micro and secondary nutrient deficiencies in soils also lead to widespread occurrence of mineral deficiency disorders such as anemia, goiter, dental caries, etc. The emerging challenges and opportunities need to be addressed and call for a paradigm shift from traditional research to innovative, demand-driven research on crucial areas for enhancing crop productivity through sustenance of soil health and fertility.

In this context, *Experimental Soil Biology and Fertility* provides a thorough introduction to practical aspects of soil fertility as well as the soil biology, and is an appropriate text for undergraduate students of agriculture, horticulture, and forestry in particular and the scientific community related to the domain of agriculture in general.

Utilizing my extensive academic and professional experience and conception of soil science, this book is written in a simple and lucid manner covering all aspects of soil biology and fertility in 41 chapters, with subheadings and supportive illustrations to assist in quickly locating topics of interest. This book is designed following the syllabi for undergraduate and postgraduate programs in various agricultural/horticultural universities, to ensure easy and clear understanding about various spheres of soil biology and fertility.

This book addresses the various exercises/experiments as laid out in the syllabus of various undergraduate courses of agriculture, horticulture, and forestry entitled "Experimental Soil Biology and Fertility," offered during undergraduate and postgraduate programs. This will serve as a guide to understand various analytical principles and instrumental methods related to soil fertility, thereby helping readers to progress in acquiring in-depth knowledge on carrying out of various research activities/programs in the field of agriculture and allied disciplines.

I am sure that this book will meet the requirement of undergraduate and postgraduate students of soil science and other disciplines related to agriculture, horticulture, and forestry, and will inspire them to expand their ideas and horizons in all possible ways.

I acknowledge the sources of figures and tables that have been reproduced from different books, journals, and online resources.

I consider it my privilege to express my deep admiration and immense gratitude to the honorable Vice Chancellor Dr. Anupam Mishra, Central Agricultural University, Imphal, for his visionary guidance and imperative efforts in aligning and contextualizing academic pursuits with contemporary and projected national and global agricultural trends, besides penning foreword for this book. The author is also very much grateful to Dr. B.N. Hazarika, Dean, College of Horticulture and Forestry, for their constant encouragement, guidance and meticulous support in bringing out of this book.

**A.S. Mailappa**
*Pasighat*

# Author Biography

**Dr. A.S. Mailappa** is a renowned soil scientist and is currently working as Associate Professor (Soil Science and Agricultural Chemistry) at the College of Horticulture and Forestry, Pasighat, Arunachal Pradesh, under Central Agricultural University, Imphal, India. He has obtained his MSc(Ag) in Soil Science and Agricultural Chemistry in 2000 from the University of Agricultural Sciences (Bengaluru) under the ICAR–Junior Research Fellowship and his PhD from Tamil Nadu Agricultural University (Coimbatore) in 2003 under the Neyveli Lignite Corporation Senior Research Fellowship. He worked as junior and senior research fellow in various projects, including the Indian Space Research Organization, from November 1998 to March 2007. He specializes in soil chemistry, soil fertility, and soil microbiology, and has published more than 22 research papers in reputed journals, 17 books, 45 book chapters, and several practical manuals and technical bulletins. He received 36 prestigious national and international awards for his immense contribution to teaching, research, and innovation including the Krishi Shikshak Award (Best Academician Award), Scientist Award, 2019 (Dr. B.V. David Foundation), Excellence in Teaching Award (SVWS), Excellence in Research Award (DRASS), Best Scientist Award (SFSN), Excellence in Innovation Award, 2020 (Dr. B.V. David Foundation), Dr. A.P.J. Abdul Kalam Scientist Award, 2020, Outstanding Teacher Award, 2020 (SBER), International Education Excellence Award in Soil Science and Agricultural Chemistry, 2021 (CPACE), Reviewer Excellence Award (ARCC), Distinguished Scientist Award, 2021 (AETDS), Dr. Rattan Lal Award, 2022 (AEEFWS), Excellence in Extension Education Award, 2022 (SBER), Entrepreneurship Development and Excellence Award, 2022 (Dr. B.V. David Foundation), Dr. V.P. Tyagi Memorial Award, 2022 (AETDS), and has guided many postgraduate and PhD students.

# Glossary[1]

**A horizons:** Mineral horizons that formed at the surface, or below an O horizon, and are characterized by an accumulation of humified organic matter intimately mixed with the mineral fraction.

**Acid soil:** Soil with a pH value < 7.0.

**Acidic cations:** Hydrogen ions or cations that, on being added to water, undergo hydrolysis, resulting in an acidic solution. Examples in soils are $H^+$, $Al^{3+}$, and $Fe^{3+}$.

**Acidity, residual:** Soil acidity that is neutralized by lime or other alkaline materials, but that cannot be replaced by an unbuffered salt solution.

**Acidity, salt-replaceable:** The aluminum and hydrogen that can be replaced from an acid soil by an unbuffered salt solution such as KCl. Essentially, the sum of the exchangeable Al and H.

**Acidity, total:** Sum of salt-replaceable and residual acidity.

**Actinomycetes:** A non-taxonomic term applied to a group of gram-positive bacteria that have a superficial resemblance to fungi.

**Activity:** Informally, may be taken as the effective concentration of a substance in a solution.

**Aerate:** To allow or promote exchange of soil gases with atmospheric gases.

**Aeration porosity:** The fraction of the bulk soil volume that is filled with air at any given time, or under a given condition, such as a specified soil-water matric potential.

**Aerobic:** (1) Having molecular oxygen as a part of the environment. (2) Growing only in the presence of molecular oxygen, as aerobic organisms. (3) Occurring only in the presence of molecular oxygen, such as aerobic decomposition.

**Aggregate:** A unit of soil structure, usually formed by natural processes, and generally below 10mm in diameter.

**Air dry:** The state of dryness at equilibrium with the water content in the surrounding atmosphere.

**Albic horizon:** A mineral soil horizon from which clay and free iron oxides have been removed or segregated. The color of the horizon is determined primarily by the

[1] Adapted from the Glossary of Soil Science Terms published by the Soil Science Society of America.

color of primary sand and silt particles rather than coatings on these particles. An E horizon.

**Alfisols:** Mineral soils that have umbric or ochric epipedons, argillic, or kandic horizons, and that have plant available water during at least 90 days when the soil is warm enough for plants to grow. Alfisols have a mean annual soil temperature below 8°C or a base saturation in the lower part of the argillic horizon of 35% or more when measured at pH 8.2. A soil order.

**Alkaline soil:** Any soil having a pH above 7.0.

**Ammonification:** The biological process leading to the formation of ammoniacal nitrogen from nitrogen containing organic compounds.

**Ammonium fixation:** The process of converting exchangeable or soluble ammonium ions to those occupying interlayer positions similar to K + in micas.

**Amorphous material:** Non-crystalline soil constituents.

**Anaerobic:** (1) The absence of molecular oxygen. (2) Growing in the absence of molecular oxygen, such as anaerobic bacteria. (3) Occurring in the absence of oxygen, such as a biochemical process.

**Andisols:** Mineral soils developed in volcanic ejecta that have andic soil properties. A soil order.

**Anion exchange capacity:** The sum total of exchangeable anions that a soil can adsorb.

**Aquic:** A mostly reducing soil moisture regime nearly free of dissolved oxygen due to saturation by groundwater or its capillary fringe and occurring at periods when the soil temperature at 50cm below the surface is above 5°C.

**Argillic horizon:** A mineral soil horizon that is characterized by the illuvial accumulation of layer-silicate clays.

**Aridic:** A soil moisture regime that has no plant available water for more than half the cumulative time that the soil temperature at 50cm below the surface is above 5°C and has no period as long as 90 consecutive days when there is water for plants, while the soil temperature at 50cm is continuously above 8°C.

**Aridisols:** Mineral soils that have an aridic moisture regime but no oxic horizon. A soil order.

**Autotroph:** An organism capable of utilizing $CO_2$ or carbonates as a sole source of carbon and obtaining energy for carbon reduction and biosynthetic processes from radiant energy (photoautotroph) or oxidation of inorganic substances (chemoautotroph).

**Available nutrients:** Nutrient ions or compounds in forms that plants can absorb and utilize in growth.

**Available water:** The portion of water in a soil that can be absorbed by plant roots; the amount of water between in situ field capacity and the wilting point.

**B horizons:** Horizons that formed below an A, E, or O horizon that have properties different from the overlying and underlying horizons owing to the soil forming processes.

**Bacteroid:** An altered form of cells of certain bacteria. Refers particularly to the swollen, irregular vacuolated cells of *Rhizobium* in nodules of legumes.

**Bar:** A unit of pressure equal to 1 million dynes per square centimeter.

**Basic cation saturation percentage:** The extent to which the cation exchange capacity is saturated with alkali (sodium and potassium) and alkaline earth (calcium and magnesium) cations expressed as a percentage of the cation exchange capacity, as measured at a particular pH, such as 7 or 8.2.

**Biomass:** The total mass of living microorganisms in a given volume or mass of soil.

**Biosequence:** A sequence of related soils, differing one from the other, primarily because of differences in kinds and numbers of plants and soil organisms as a soil-forming factor.

**Biuret:** A toxic product formed at high temperature during the manufacturing of urea.

**Buffer power:** The ability of ions associated with the solid phase to buffer changes in ion concentration in the solution phase.

**Bulk density, soil:** The mass of dry soil per unit bulk volume, expressed as grams per cubic centimeter, $g/cm^3$. The bulk volume is determined before drying to constant weight at 105°C.

**C horizons:** Horizons or layers, excluding hard rock, that are little affected by the soil-forming processes. C horizons typically underlying A, B, E, or O horizons.

**Calcareous soil:** Soil containing sufficient free $CaCO_3$, and/or $MgCO_3$, to effervesce visibly when treated with cold 0.1 molar HCl. These soils usually contain from as little as 1–20% $CaCO_3$ equivalent.

**Calcic horizon:** A mineral soil horizon of secondary carbonate enrichment that is more than 15cm thick, has a $CaCO_3$ equivalent above 150g/kg, and has at least 50g/kg more $CaCO_3$ equivalent than the underlying C horizon.

**Caliche:** A zone near the surface, more or less cemented by secondary carbonates of calcium or magnesium precipitated from the soil solution. Caliche may occur as a soft thin soil horizon, as a hard thick bed, or as a surface layer exposed by erosion.

**Cambic horizon:** A mineral soil horizon that has a texture of loamy very fine sand or finer, has soil structure rather than rock structure, contains some weatherable minerals, and is characterized by alteration or removal of mineral material, or the removal of carbonates.

**Capillary fringe:** A zone in the soil just above the water table that remains saturated or almost saturated with water.

**Carbon-organic nitrogen ratio:** The ratio of the mass of organic carbon to the mass of organic nitrogen in soil, organic material, plants, or the cells of microorganisms.

**Cat clay:** Wet clay soils containing ferrous sulfide, which become highly acidic when drained.

**Catena:** A sequence of soils of about the same age, derived from similar parent material, and occurring under similar climatic conditions, but having different characteristics because of variations in relief and in drainage.

**Cation exchange:** The interchange between a cation in solution and another cation on the surface of any negatively charged material such as clay colloid or organic colloid.

**Cation exchange capacity (CEC):** The sum of exchangeable cations that a soil, soil constituent, or other material can adsorb at a specific pH; commonly expressed as milliequivalents per 100g or centimoles per kilogram.

**Chemical potential:** Informally, it is the capacity of a solution or other substance to do work by virtue of its chemical composition.

**Chlorosis:** Failure of plants to develop chlorophyll caused by a deficiency of an essential element. Chlorotic leaves range in color from light green through yellow to almost white.

**Chroma:** The relative purity, strength, or saturation of a color; directly related to the dominance of the determining wavelength of the light and inversely related to grayness; one of the three variables of color.

**Chronosequence:** A sequence of related soils that differ one from the other, in certain properties primarily as a result of time as a soil-forming factor.

**Citrate-soluble phosphorus:** That part of the total phosphorus in fertilizer that is insoluble in water but soluble in neutral 0.33M ammonium citrate and which, together with water-soluble phosphorus represents the readily available phosphorus content of the fertilizer.

**Clay:** (1) A soil separate consisting of particles.
(2) A textural class.

**Clay films:** Coatings of clay on the surfaces of soil peds, mineral grains, and in soil pores. (Also called clay skins, clay lows, or argillans.)

**Clay mineral:** Any crystalline inorganic substance of clay size.

**Claypan:** A dense, compact layer in the subsoil having a much higher clay content than the overlying material, from which it is separated by a sharply defined boundary. Claypans usually impede the movement of water and air and the growth of plant roots.

**Colluvium:** A general term applied to deposits on a slope or at the foot of a slope or cliff that were moved there chiefly by gravity.

**Concretion:** A local concentration of a chemical compound, such as calcium carbonate or iron oxide, in the form of a grain or nodule of varying size, shape, hardness, and color.

**Consistency:** The manifestations of the forces of cohesion and adhesion acting within the soil at various water contents, as expressed by the relative ease with which a soil can be deformed or ruptured.

**Consumptive water use:** The water used by plants in transpiration and growth, plus water vapor loss from adjacent soil or snow, or from intercepted precipitation in any specified time. Usually expressed as equivalent depth of free water per unit of time.

**Creep:** Slow mass movement of soil and soil material down relatively steep slopes primarily under the influence of gravity.

**Crust:** A soil-surface layer, ranging in thickness from a few millimeters to a few tens of millimeters, which, when dry, is much more compact, hard, and brittle than the material immediately beneath it.

**Cryic:** A soil temperature regime that has mean annual soil temperatures of above 0°C but below 8°C at 50cm. There is more than a 5°C difference between mean summer and mean winter soil temperatures.

**Darcy's law:** A law describing the rate of flow of water through porous media. Named for Henry Darcy of Paris, who formulated it in 1856 from extensive work on the flow of water through sand filter beds.

**Denitrification:** Reduction of nitrate or nitrite to molecular nitrogen or nitrogen oxides by microbial activity or by chemical reactions involving nitrite.

**Desert pavement:** The layer of gravel or stones remaining on the land surface in desert regions after the removal of the fine material by wind erosion.

**Diffuse double layer:** A heterogeneous system that consists of a solid surface layer having a net electrical charge, together with an ionic swarm under the influence of the solid and a solution phase that is in direct contact with the surface.

**Duripan:** A mineral soil horizon that is cemented by silica to the point that air-dry fragments will not slake in water or HCl.

**E horizons:** Mineral horizons in which the main feature is loss of silicate clay, iron, aluminum, or some combination of these, leaving a concentration of sand and silt particles of quartz or other resistant minerals.

**ECe:** The electrolytic conductivity of an extract from saturated soil, normally expressed in units of decisiemens per meter at 25°C.

**Ecology:** The science that deals with the interrelationships between organisms and between organisms and their environment.

**Ecosystem:** A community of organisms and the environment in which they live.

**Ectomycorrhiza:** A mycorrhizal association in which the fungal mycelia extend inward, between root cortical cells, to form a network ("Hartig net") and outward into the surrounding soil.

**Edaphology:** The science that deals with the influence of soils on living things, particularly plants.

**Eluviation:** The removal of soil material in suspension (or in solution) from a layer or layers of soil.

**Endomycorrhiza:** A mycorrhizal association with intracellular penetration of the host root cortical cells by the fungus as well as outward extension into the surrounding soil.

**Entisols:** Mineral soils that have no distinct subsurface diagnostic horizons within 1m of the soil surface. A soil order.

**Erosion:** (1) The wearing away of the land surface by running water, wind, ice, and other geological agents, including such processes as gravitational creep. (2) Detachment and movement of soil or rock by water, wind, ice, or gravity.

**Erosion potential:** A numerical value expressing the inherent erodibility of a soil.

**Eutrophic:** Having concentrations of nutrients optimal, or nearly so, for plant or animal growth.

**Evapotranspiration:** The combined loss of water from a given area, and during a specified period of time, by evaporation from the soil surface and by transpiration from plants.

**Exchangeable cation percentage:** The extent to which the adsorption complex of a soil is occupied to particular cation.

**Exchangeable ion:** A cation or anion held on or near the surface of a solid particle, which may be replaced by other ions of similar charge that are in solution.

**Fallowing:** The practice of leaving land uncropped for periods of time to accumulate and retain water and mineralized nutrient elements.

**Family, soil:** In soil classification one of the categories intermediate between the great soil group and the soil series. Families provide groupings of soils with ranges in texture, mineralogy, temperature, and thickness.

**Fertigation:** Application of plant nutrients in irrigation water.

**Fertility, soil:** The ability of a soil to supply elements essential for plant growth without a toxic concentration of any element.

**Fertilizer:** Any organic or inorganic material of natural or synthetic origin (other than liming materials) that is added to a soil to supply one or more elements essential to the growth of plants.

**Fibric material:** Mostly undecomposed plant remains that contain large amounts of well-preserved and recognizable fibers.

**Field capacity, in situ (field water capacity):** The content of water, on a mass or volume basis, remaining in a soil two or three days after having been wetted with water and after free drainage is negligible.

**Film water:** A thin layer of water, in close proximity to soil-particle surfaces, that varies in thickness from one or two to perhaps 100 or more molecular layers.

**Fixation:** The process by which available plant nutrients are rendered less available or unavailable in the soil.

**Floodplain:** The land bordering a stream, built up of sediments from overflow of the stream, and subject to inundation when the stream is at flood stage.

**Flux:** Thc timc ratc of transport of a quantity across a given area.

**Free iron oxides:** A general term for those iron oxides that can be reduced and dissolved by a dithionite treatment. Often includes goethite and hematite.

**Friable:** A consistency term pertaining to the ease of crumbling of soils.

**Fluvic acid:** The colored material that remains in solution after the removal of humic acid by acidification.

**Gibbsite:** A mineral with a platy habit that occurs in highly weathered soils. $Al(OH)_3$.

**Gilgai:** The microrelief of soils produced by expansion and contraction with changes in water content, a common feature of vertisols.

**Glacial drift:** Rock debris that has been transported by glaciers and deposited, either directly from the ice or from the meltwater.

**Goethite:** A yellow-brown iron oxide mineral that is very common and is responsible for the brown color in many soils. FeOOH.

**Great soil group:** One of the categories in the system of soil classification that has been used in the United

States for many years. Great groups place soils according to soil moisture and temperature, basic cation saturation status, and expression of soil horizons.

**Green manure:** Plant material incorporated into soil while green or at maturity; for soil improvement.

**Groundwater:** That portion of the water below the surface of the ground at a pressure equal to or greater than atmospheric.

**Guano:** The decomposed dried excrement of birds and bats, used for fertilizer purposes.

**Gypsic horizon:** A mineral soil horizon of secondary calcium sulfate enrichment that is more than 15cm thick.

**Gypsum:** The common name for calcium sulfate ($CaSO_4.2H_2O$), used to supply calcium and sulfur to ameliorate sodic soils.

**Hardpan:** A hardened soil layer, in the lower A or in the B horizon, caused by cementation of soil particles with organic matter or with materials such as silica, sesquioxides, or calcium carbonate.

**Heavy metals:** Those metals that have high density; in agronomic usage includes Cu, Fe, Mn, Mo, Co, Zn, Cd, Hg, Ni, and Pb.

**Hematite:** A red iron oxide mineral that contributes red color to many soils. $Fe_2O_3$.

**Hemic material:** An intermediate degree of decomposition, such as two-thirds of the organic material cannot be recognized.

**Heterotroph:** An organism capable of deriving carbon and energy for growth and cell synthesis by the utilization of organic compounds.

**Histosols:** Organic soils that have organic soil material in more than half of the upper 80cm, or that are of any thickness overlying rock or fragmental materials, which have interstices filled with organic soil materials. A soil order.

**Hue:** One of the three variables of color.

**Humic acid:** The dark-colored organic material that can be extracted from soil by various agents and that is precipitated by acidification to pH 1 or 2.

**Humification:** The process whereby the carbon of organic residues is transformed and converted to humic substances through biochemical and/or chemical processes.

**Humus:** All of the organic compounds in soil exclusive of undecayed plant and animal tissues, their partial decomposition products, and the soil biomass. Resistant to further alteration.

**Hydraulic conductivity:** The proportionality factor in Darcy's law as applied to the viscous flow of water in soil.

**Hydroxy-aluminum:** Aluminum hydroxide compounds of varying composition.

**Hyperthermic:** A soil temperature regime that has mean annual soil temperatures of 22°C or more and a higher than 5°C difference between mean summer and mean winter soil temperatures at 50cm below the surface.

**Illuvial horizon:** A soil layer or horizon in which material carried from the overlying layer has been precipitated from solution or deposited from suspension. The layer of accumulation.

**Illuviation:** The process of deposition of soil material removed from one horizon to another in the soil, usually from an upper to a lower horizon in the soil profile.

**Immobilization:** The conversion of an element from the inorganic to the organic form in microbial or plant tissues.

**Inceptisols:** Mineral soils having one or more pedogenic horizons in which mineral materials, other than carbonates or amorphous silica, have been altered or removed but not accumulated to a significant degree. Water is available to plants more than half of the year or more than 90 consecutive days during a warm season. A soil order.

**Indicator plants:** Plants characteristic of specific soil or site conditions.

**Infiltration:** The downward entry of water through the soil surface.

**Ion activity:** Informally, the effective concentration of an ion in solution.

**Iron oxides:** Group name for the oxides and hydroxides of iron. Includes the minerals goethite, hematite, lepidocrocite, ferrihydrite, maghemite, and magnetite. Sometimes referred to as sesquioxides or hydrous oxides.

**Iron pan:** An indurated soil horizon in which iron oxide is the principal cementing agent, as in plinthite or laterite.

**Ironstone:** Hardened plinthite materials often occurring as nodules and concretions.

**Isomorphous substitution:** The replacement of one atom by another of similar size in a crystal structure without disrupting or seriously changing the structure.

**Jarosite:** A yellow potassium iron sulfate mineral.

**Kandic horizon:** Subsoil diagnostic horizon having a clay increase relative to overlying horizons and having low-activity clays, below 16meq/100g or below 16cmol(+)/kg of clay.

**Kaolinite:** A clay mineral of the kaolin subgroup. It has a 1:1 layer structure and is a non-expanding clay mineral.

**Labile:** A substance that is readily transformed by microorganisms or is readily available to plants.

**Landscape:** All the natural features such as fields, hills, forests, water, and such, which distinguish one part of the earth's surface from another part. Usually that portion of the land that the eye can comprehend in a single view.

**Lattice:** A regular geometric arrangement of points in a plane or in space. A lattice is used to represent the distribution of repeating atoms or groups of atoms in a crystalline substance.

**Leaching:** The removal of materials in solution from the soil.

**Lepidocrocite:** An orange iron oxide mineral that is found in mottles and concretions of wet soils. FeOOH.

**Lime, agricultural:** A soil amendment containing calcium carbonate, magnesium carbonate, and other materials; used to neutralize soil acidity and furnish calcium and magnesium for plant growth.

**Lithosequence:** A group of related soils that differ, one from the other, in certain properties; primarily a result of differences in the parent rock as a soil-forming factor.

**Loam:** A soil textural class.

**Loess:** Material transported and deposited by wind and consisting of predominantly silt-sized particles.

**Luxury uptake:** The absorption by plants of nutrients in excess of their need for growth.

**Macronutrient:** A plant nutrient usually attaining a concentration of more than 500mg/kg in mature plants.

**Maghemite:** A dark, reddish-brown, magnetic iron oxide mineral chemically similar to hematite, but structurally similar to magnetite. $Fe_2O_3$. Often found in well-drained, highly weathered soils of the tropical regions.

**Magnetite:** A black, magnetic iron oxide mineral usually inherited from igneous rocks. Often found in soils as black magnetic sand grains.

**Manganese oxides:** A group term for oxides of manganese. They are typically black and frequently occur as nodules and coatings on ped faces, usually in association with iron oxides.

**Manure:** The excreta of animals, with or without an admixture of bedding or litter, fresh or at various stages of further decomposition or composting.

**Mass flow (nutrient) :** The movement of solutes associated with the net movement of water.

**Melanic horizon:** A thick, dark-colored surface horizon having andic soil properties.

**Mesic:** A soil temperature regime that has mean annual soil temperatures of 8°C or more but less than 15°C, and more than 5°C difference between mean summer and mean winter temperatures at 50cm below the surface.

**Mica:** A layer-structured aluminosilicate mineral group of the 2:1 type that is characterized by high layer charge, which is usually satisfied by potassium.

**Microclimate:** The sequence of atmospheric changes within a very small region.

**Microfauna:** Protozoa, nematodes, and arthropods of microscopic size.

**Microflora:** Bacteria (including actinomycetes), fungi, algae, and viruses.

**Micronutrient:** A chemical element necessary for plant growth found in small amounts, usually less than 100mg/kg in the plant. These elements consist of B, Cl, Cu, Fe, Mn, Mo, and Zn.

**Mineralization:** The conversion of an element from an organic form to an inorganic state as a result of microbial activity.

**Mineral soil:** A soil consisting predominantly of, and having properties determined predominantly by, mineral matter. Usually contains less than 200g/kg of organic carbon.

**Mollisols:** Mineral soils that have a mollic epipedon overlying mineral material with a basic cation saturation of 50% or more when measured at pH 7.0. A soil order.

**Montmorillonite:** An aluminum silicate (smectite) with a layer structure composed of two silica tetrahedral sheets and a shared aluminum and magnesium octahedral sheet.

**Mor:** A type of forest humus in which the Oa horizon is present and in which there is almost no mixing of surface organic matter with mineral soil.

**Mottled zone:** A layer that is marked with spots or blotches of different color or shades of color (mottles).

**Muck soil:** An organic soil in which the plant residues have been altered beyond recognition.

**Mulch:** Any material such as straw, sawdust, leaves, plastic film, and loose soil, that is spread upon the soil surface to protect soil and plant roots from the effects of raindrops, soil crusting, freezing, evaporation, and such.

**Mulch farming:** A system of tillage and planting operations resulting in minimum incorporation of plant residues or other mulch into the soil surface.

**Mull:** A type of forest humus in which the Oe horizon may or may not be present and in which there is no Oa horizon. The A horizon consists of an intimate mixture of organic matter and mineral soil with gradual transition between the A horizon and the horizon underneath.

**Munsell color system:** A color designation system that specifies the relative degrees of the three simple variables of color: hue, value, and chroma. For example: 10YR 6/4 is a color with a hue = 10YR (yellow–red), value = 6, and chroma = 4.

**Mycorrhiza:** Literally "fungus root." The association, usually symbiotic, of specific fungi with the roots of higher plants.

**N value:** The relationship between the percentage of water under field conditions and the percentages of inorganic clay and of humus.

**Natric horizon:** A mineral soil horizon that satisfies the requirements of an argillic horizon, but that also has prismatic, columnar, or blocky structure and a subhorizon having 15% or more saturation with exchangeable sodium.

**Nitrification:** Biological oxidation of ammonium to nitrite and nitrate, or a biologically induced increase in the oxidation state of nitrogen.

**Nitrogenase:** The specific enzyme required for biological dinitrogen fixation.

**No-tillage system:** A procedure whereby a crop is planted directly into the soil with no preparatory tillage since harvest of the previous crop; usually a special planter is necessary to prepare a narrow, shallow seedbed immediately surrounding the seed being planted.

**Nutrient antagonism:** The depressing effect caused by one or more plant nutrients on the uptake and availability of another.

**Nutrient interaction:** A statistical term used when two or more nutrients are applied together to denote a departure from additive responses occurring when they are applied separately.

**O horizon:** Layers dominated by organic material, except limnic layers that are organic.

**Ochric epipedon:** A thin, light-colored surface horizon of mineral soil.

**Organic soil:** A soil that contains a high percentage of organic carbon (>200g/kg or >120–180g/kg if saturated with water) throughout the solum.

**Outwash:** Stratified glacial drift deposited by meltwater streams beyond active glacier ice.

**Oven-dry soil:** Soil that has been dried at 105°C until it reaches constant mass.

**Oxisols:** Mineral soils that have an oxic horizon within 2m of the surface or plinthite as a continuous phase within 30cm of the surface, and that do not have a spodic or argillic horizon above the oxic horizon. A soil order.

**Paleosol, buried:** A soil formed on a landscape during the geological past and subsequently buried by sedimentation.

**Pans:** Horizons or layers in soils that are strongly compacted, indurated, or having very high clay content.

**Paralithic contact:** Similar to lithic contact except that it is softer, and is difficult to dig with a spade.

**Parent material:** The unconsolidated and more or less chemically weathered mineral or organic matter from which the solum of soils is developed by pedogenic processes.

**Particle density:** The density of soil particles, the dry mass of the particles being divided by the solid volume of the particles.

**Pascal:** A unit of pressure equal to 1 Newton per square meter.

**Peat:** Unconsolidated soil material consisting largely of undecomposed, or only slightly decomposed, organic matter accumulated under conditions of excessive moisture.

**Ped:** A unit of soil structure such as an aggregate, crumb, prism, block, or granule, formed by natural processes.

**Pedon:** A three-dimensional body of soil with lateral dimensions large enough to permit the study of horizon shapes and relations. Its area ranges from $1m^2$ to $10m^2$.

**Penetrability:** The ease with which a probe can be pushed into the soil.

**Percolation:** The downward movement of water through the soil.

**Pergelic:** A soil temperature regime that has mean annual soil temperatures of less than 0°C. Permafrost is present.

**Permafrost:** A perennially frozen soil horizon.

**Permafrost table:** The upper boundary of the permafrost, coincident with the lower limit of seasonal thaw.

**Permanent wilting point:** The largest water content of a soil at which indicator plants, growing in that soil, wilt and fail to recover when placed in a humid chamber. Often estimated by the water content at −15 bars, −1500 kilopascals, or −1.5 megapascals soil matric potential.

**Petrocalcic horizon:** A continuous, indurated calcic horizon that is cemented by calcium carbonate and, in some places, with magnesium carbonate.

**Petrogypsic horizon:** A continuous, strongly cemented, massive gypsic horizon that is cemented with calcium sulfate.

**pH-dependent charge:** The portion of the cation or anion exchange capacity which varies with pH.

**pH, soil:** The negative logarithm of the hydrogen ion activity of a soil.

**Phase:** A utilitarian grouping of soils defined by soil or environmental features that are not class differentia used in the US system of soil taxonomy, for example, surface texture, surficial rock fragments, salinity, erosion, thickness, and such.

**Phosphate:** In fertilizer trade terminology, phosphate is used to express the sum of the water-soluble and citrate-soluble phosphoric acid ($P_2O_5$); also referred to as the available phosphoric acid ($P_2O_5$)

**Phosphoric acid:** In commercial fertilizer manufacturing, it is used to designate orthophosphoric acid, $H_3PO_4$. In fertilizer labeling, it is the common term used to represent the phosphate content in terms of available phosphorus, expressed as percent $P_2O_5$.

**$P_2O_5$:** Phosphorus pentoxide; fertilizer label designation that denotes the percentage of available phosphate.

**Placic horizon:** A black to dark-reddish mineral soil horizon that is usually thin, is commonly cemented with iron, and is slowly permeable or impenetrable to water and roots.

**Plastic soil:** A soil capable of being molded or deformed continuously and permanently, by relatively moderate pressure, into various shapes. See **consistency**.

**Plinthite:** A non-indurated mixture of iron and aluminum oxides, clay, quartz, and other diluents that commonly occurs as red soil mottles usually arranged in platy, polygonal, or reticulate patterns. Plinthite changes irreversibly to ironstone hardpans, or irregular aggregates on exposure to repeated wetting and drying.

**Plow pan:** An induced subsurface soil horizon or layer having a higher bulk density and lower total porosity than the soil material directly above and below, but similar in particle size analysis and chemical properties. The pan is usually found just below the maximum depth of primary tillage and frequently restricts root development and water movement. Also called a pressure pan or plough sole.

**Polypedon:** A group of contiguous similar pedons.

**Potash:** Term used to refer to potassium or potassium fertilizers and usually designated as $K_2O$.

**Potassium fixation:** The process of converting exchangeable or water-soluble potassium to that occupying the position of $K^+$ in the micas.

**Primary mineral:** A mineral that has not been altered chemically since deposition and crystallization from molten lava.

**Profile, soil:** A vertical section of soil through all its horizons and extending into the C horizon.

**R layer:** Hard bedrock including granite, basalt, quartzite, and indurated limestone or sandstone that is sufficiently coherent to make hand digging impractical.

**Rainfall erosion index:** A measure of the erosive potential of a specific rainfall event.

**Reaction, soil:** The degree of acidity or alkalinity of a soil, usually expressed as a pH value.

**Regolith:** The unconsolidated mantle of weathered rock and soil material on the earth's surface; loose earth materials above solid rock.

**Residual fertility:** The available nutrient content of a soil carried to subsequent crops.

**Reticulate mottling:** A network of streaks of different color, most commonly found in the deeper profiles of soil containing plinthite.

**Rhizobia:** Bacteria capable of living symbiotically in roots of leguminous plants, from which they receive energy and often utilize molecular nitrogen.

**Rhizosphere:** The zone immediately adjacent to plant roots in which the kinds, numbers, or activities of microorganisms differ from that of the bulk soil.

**Rill:** A small, intermittent watercourse with steep sides; usually only several centimeters deep and, thus, no obstacle to tillage operations.

**Runoff:** That portion of the precipitation on an area which is discharged from the area through stream channels.

**Saline soil:** A non-sodic soil containing sufficient soluble salt to adversely affect the growth of most crop plants.

**Salic horizon:** A mineral soil horizon of enrichment with secondary salts more soluble in cold water than gypsum.

**Saline seep:** Intermittent or continuous saline water discharge at or near the soil surface under dry-land conditions, which reduces or eliminates crop growth.

**Saline-sodic soil:** A soil containing both sufficient soluble salt and exchangeable sodium to adversely affect crop production under most soil and crop conditions.

**Salt balance:** The quantity of soluble salt removed from an irrigated area in the drainage water minus that delivered in the irrigation water.

**Sand:** (1) A soil particle between 0.05mm and 2.0mm in diameter. (2) A soil textural class.

**Sapric material:** One of the components of organic soils with highly decomposed plant remains. Material is not recognizable and bulk density is low.

**Saprolite:** Weathered rock materials that maybe soil parent material.

**Saturation extract:** The solution extracted from a soil at its saturation water content.

**Secondary mineral:** A mineral resulting from the decomposition of a primary mineral or from the reprecipitation of the products of decomposition of a primary mineral.

**Self-mulching soil:** A soil in which the surface layer becomes so well aggregated that it does not crust and seal under the impact of rain, but instead serves as a surface mulch upon drying.

**Series, soil:** See **soil series**.

**Sesquioxides:** A general term for oxides and hydroxides of iron and aluminum.

**Siderophore:** A non-porphyrin metabolite secreted by certain microorganisms and plant roots that forms a highly stable coordination compound with iron.

**Silica-alumina ratio:** The molecules of silicon dioxide per molecule of aluminum oxide in clay minerals or in soils.

**Silt:** (1) A soil separate consisting of particles between 0.05mm and 0.002mm in equivalent diameter. (2) A soil textural class.

**Site index:** A quantitative evaluation of the productivity of a soil for forest growth under the existing or specified environment. Commonly, the height in meters of the dominant forest vegetation taken or calculated to an age index such as 25, 50, or 100 years.

**Slickensides:** Polished and grooved surfaces produced by one mass sliding past another; common in vertisols.

**Smectite:** A group of 2:1-layer structured silicates with high cation exchange capacity and variable interlayer spacing.

**Sodic soil:** A non-saline soil containing sufficient exchangeable sodium to adversely affect crop production and soil structure under most conditions of soil and plant type.

**Sodium adsorption ratio (SAR):** A relation between soluble sodium and soluble divalent cations that can be used to predict the exchangeable sodium percentage of soil equilibrated with a given solution. Defined as: where concentrations, denoted by parentheses, are expressed in moles per liter.

**Soil:** (1) The unconsolidated mineral or material on the immediate surface of the earth that serves as a natural medium for the growth of land plants. (2) The unconsolidated mineral or organic matter on the surface of the earth, which has been subjected to and influenced by genetic and environmental factors of parent material, climate, macro- and microorganisms, and topography, all acting over a period of time and producing a product—soil—that differs from the material from which it is derived in many physical, chemical, biological, and morphological properties and characteristics.

**Soil association:** A kind of map unit used in soil surveys comprised of delineations, each of which shows the size, shape, and location of a landscape unit composed of two or more kinds of component soils, or component soils and miscellaneous areas, plus allowable inclusions in either case.

**Soil conservation:** A combination of all management and land-use methods that safeguard the soil against depletion or deterioration by natural or human-induced factors.

**Soil genesis:** The mode of origin of the soil with special reference to the processes or soil-forming factors responsible for development of the solum, or true soil, from unconsolidated parent material.

**Soil horizon:** A layer of soil or soil material approximately parallel to the land surface and differing from adjacent genetically related layers in physical, chemical, and biological properties or characteristics such as color, structure, texture, consistency, kinds and number of organisms present, degree of acidity or alkalinity, and so on.

**Soil loss tolerance:** (1) The maximum average annual soil loss that will allow continuous cropping and maintain soil productivity without requiring additional management inputs. (2) The maximum soil erosion loss that is offset by the theoretical maximum rate of soil development, which will maintain an equilibrium between soil losses and gains.

**Soil management groups:** Groups of taxonomic soil units with similar adaptations or management requirements for one or more specific purposes, such as adapted crops or crop rotations, drainage practices, fertilization, forestry, and highway engineering.

**Soil monolith:** A vertical section of a soil profile removed from the soil and mounted for display or study.

**Soil productivity:** The capacity of a soil to produce a certain yield of crops, or other plants, with optimum management.

**Soil science:** That science dealing with soils as a natural resource on the surface of the earth, including soil formation, classification and mapping, geography and use, and physical, chemical, biological, and fertility properties of soils per se: and those properties in relation to their use and management.

**Soil separates:** Mineral particles, less than 2.0mm in equivalent diameter, ranging between specified size limits.

**Soil series:** The lowest category in the US system of soil taxonomy: a conceptualized class of soil bodies (polypedons) that have limits and ranges more restrictive than all higher taxa. The soil series serves as a major vehicle to transfer soil information and research knowledge from one soil area to another.

**Soil solution:** The aqueous liquid phase of the soil and its solutes.

**Soil structure:** The combination or arrangement of primary soil particles into secondary particles, units, or peds.

**Soil survey:** The systematic examination, description, classification, and mapping of soils in an area.

**Soil texture:** The relative proportions of the various soil separates in a soil.

**Soil water potential (total) :** The amount of work that must be done per unit quantity of pure water in order to transport reversibly and isothermally an infinitesimal quantity of water from a pool of pure water, at a specified elevation and at atmospheric pressure, to the soil water (at the point under consideration).

**Solum:** The upper and most weathered part of the soil profile; the A, E, and B horizons.

**Spodic horizon:** A mineral soil horizon that is characterized by the illuvial accumulation of amorphous materials composed of aluminum and organic carbon with or without iron.

**Spodosols:** Mineral soils that have a spodic or a placic horizon that overlies a fragipan. A soil order.

**Strip cropping:** The practice of growing crops that require different types of tillage, such as row and sod, in alternate strips along contours or across the prevailing direction of the wind.

**Structural charge:** The charge (usually negative) on a mineral caused by isomorphous substitution within the mineral layer.

**Subsoiling:** Any treatment to loosen soil below the tillage zone without inversion and with a minimum of mixing with the tilled zone.

**Surface charge density:** The excess of negative or positive charge per unit area of surface area of soil or soil mineral.

**Thermic:** A soil temperature regime that has mean annual soil temperatures of 15°C or more, but less than 22°C and more than 5°C difference between mean summer and mean winter soil temperatures at a 50cm depth below the surface.

**Thermophile:** An organism that grows readily at temperatures above 45°C.

**Till:** (1) Unstratified glacial drift deposited by ice and consisting of clay, silt, sand, gravel, and boulders, intermingled in any proportion. (2) To prepare the soil for seeding; to seed or cultivate the soil.

**Tilth:** The physical condition of soil as related to its ease of tillage, fitness as a seedbed, and its impedance to seedling emergence and root penetration.

**Top dressing:** An application of fertilizer to a soil surface, without incorporation, after the crop stand has been established.

**Toposequence:** A sequence of related soils. The soils differ, one from the other, primarily because of topography as a soil-formation factor.

**Torric:** A soil-moisture regime defined like aridic moisture regime but used in a different category of the US soil taxonomy.

**Truncated:** Having lost all or part of the upper soil horizon or horizons.

**Tuff:** Volcanic ash usually more or less stratified and in various states of consolidation.

**Udic:** A soil moisture regime that is neither dry for as long as 90 cumulative days nor for as long as 60 consecutive days in the 90 days following the summer solstice at periods when soil temperature at 50cm below the surface is above 5°C.

**Ultimate particles:** Individual soil particles after a standard dispersing treatment.

**Ultisols:** Mineral soils that have an argillic or kandic horizon with a basic cation saturation of less than 35% when measured at pH 8.2. Ultisols have a mean annual soil temperature of 8°C or higher. A soil order.

**Umbric epipedon:** A surface layer of mineral soil that has the same requirements as the mollic epipedon with respect to color, thickness, organic carbon content, consistence, structure, and phosphorus content, but that has a basic cation saturation less than 50% when measured at pH 7.

**Ustic:** A soil moisture regime that is intermediate between the aridic and udic regimes and common in temperate subhumid or semiarid regions, or in tropical and subtropical regions with a monsoon climate. A limited amount of water is available for plants but occurs at times when the soil temperature is optimum for plant growth.

**Value, color:** The relative lightness or intensity of color and approximately a function of the square root of the total amount of light. One of the three variables of color.

**Vermiculite.** A highly charged layer-structured silicate of the 2:1 type that is formed from mica.

**Vertisols:** Mineral soils that have 30% or more clay, deep wide cracks when dry, and either gilgai microrelief, intersecting slickensides, or wedge-shaped structural aggregates tilted at an angle from the horizon. A soil order.

**Vesicular arbuscular:** A common endomycorrhizal association produced by phycomycetous fungi of the family Endogonaceae. Host range includes most agricultural and horticultural crops.

**Waterlogged:** Saturated or nearly saturated with water.

**Water potential:** See **soil water potential**.

**Water table:** The upper surface of groundwater or that level in the ground where the water is at atmospheric pressure.

**Water table, perched:** The water table of a saturated layer of soil that is separated from an underlying saturated layer by an unsaturated layer (vadose water).

**Wilting point:** See **permanent wilting point**.

**Xeric:** A soil moisture regime common to Mediterranean climates having moist, cool winters and warm, dry summers. A limited amount of water is present but does not occur at optimum periods for plant growth. Irrigation or summer fallow is commonly necessary for crop production.

**Xerophytes:** Plants that grow in or on extremely dry soils.

**Yield:** The amount of a specified substance produced (e.g., grain, straw, total dry matter) per unit area.

**Zero point of charge:** The pH value of a solution in equilibrium with a particle whose net charge from all sources is zero.

**Zero tillage:** See **no-tillage system**.

*Adapted from Glossary of Soil Science Terms, published by the Soil Science Society of America.

# 1 Good Laboratory Practices and Laboratory Safety

*OECD (1998)*

## 1.1 INTRODUCTION

There are certain basic fundamental principles and practices of laboratory work at various steps of experimentation such as handling, processing, and analysis of samples and dos and don'ts associated thereof in the laboratory. It is necessary to understand that the laboratories should maintained as healthy working environments. The optimal conditions for a safe and healthy working place must be ensured regarding the presence of volatile hazardous organic liquids, explosive reagents, etc. This is made possible by following prescribed safety standards. Some common rules/regulations must be followed by the chemists and technicians in the laboratories to ensure a safe, healthy environment. These are known as Good Laboratory Practices (GLP), as detailed below.

## 1.2 SAFETY ASPECTS

- Do not keep more chemicals than the required quantity for ongoing laboratory work. The rest shall be stored in a safe place.
- Ensure safety shoes, laboratory coats, and safety glasses are worn.
- Eating, drinking, and smoking are strictly prohibited in the laboratory.
- All laboratory staff should be well acquainted with the proper functioning of all the safety equipment and personal protective equipment/devices.
- All safety devices and manuals must be kept in a place where they can be easily accessed.
- All fire extinguishers, fume hoods, chemical spill kits, eye washes, and other safety devices must be stored in appropriate places. A fully stocked first-aid kit must always be ready in the laboratory.
- Follow the safety procedure precisely for the disposal of used reagents, liquids, or chemicals.
- No one should work alone in the laboratory.
- Check and confirm the analytical procedures if they are not clear, especially while working with the dangerous organic liquids.
- Handle carefully the power supply, gas cylinders, and heating equipment.

DOI: 10.1201/9781003430100-1

- Try to work as much as possible in a fume hood and always add acid or base to water.
- Do not attempt to catch falling glassware. It is better to remove and replace broken glassware.
- New laboratory workers should be fully briefed regarding the safety precautions.
- Laboratory staff must undergo an annual medical examination.
- Standard safety procedures should be displayed as a poster in the laboratory.
- Suitable hands-on training for laboratory staff must be provided to handle emergencies.

## 1.3 LABORATORY HAZARDS

- Contact with reagents sometimes may cause external or internal injuries. External injuries are caused by skin exposure to caustic/corrosive reagents (acid/base/reactive salts). Prevention is better than cure. It is better to prevent as far as possible the spillage and splashes of chemicals/regents.
- Internal injuries may result from the toxic and corrosive effects of reagents accidently ingested and absorbed by body (for example, during manual pipetting).
- All inorganic acids and bases have their own health and safety limits. The exposure to fumes may lead to irritation or damage of eyes or skin, and could also pose respiratory issues. Hot acids are capable of reacting quickly with the skin.
- Store acids and bases separately, in well-ventilated areas, and away from volatile organic and oxidizable substances.
- Add strong acids and bases to water slowly to avoid spattering. Do not attempt the reverse process. If there is accidental skin contact with any reagents (acids, bases, toxic chemicals), thoroughly flush the contaminated area with water and seek immediate medical attention.
- Perchloric acid reacts violently or explosively on contact with the organic materials. Therefore, do not use perchloric acid together with any organic reagents, particularly volatile solvents, in one fume hood.
- Some metals (arsenic, nickel, mercury) are highly toxic and may also be carcinogenic. Avoid the inhalation, ingestion and skin contact of such chemicals.
- Sodium hydroxide (NaOH) and certain other regents produce considerable heat upon dissolution, which may lead to burns.
- Almost all organic solvents are hazardous in nature. Some are probably carcinogenic and should be handled with extra caution.
- Avoid mouth or manual pipetting. Use of VacuPads, SteriPettes, or automated pipettes is highly advisable.
- Be cautious regarding the physical hazards from electrical items and gas cylinders.
- It is not advisable to use glassware for HF (hydrofluoric acid) treatment.
- Handle hot glassware carefully, as it looks exactly like cold glassware.

## 1.4 PREPARING FOR THE LAB

It is necessary to provide hands-on training to students regarding the principles and important techniques for safe handling of reagents, instruments, etc. To achieve the above, a few of the most important suggestions are listed below.

- Don't forget to read the laboratory experiment and any suggested additional reading before coming to the laboratory.
- Find a list of questions regarding the experiments and get it clarified before proceeding to work. This will help to save your time and increase work efficiency.
- The experimental data should be recorded in an appropriate log-book or observation note. Use of loose sheets or a spiral notebook is to be avoided, as the loose sheets can be easily lost.
- Always clean your glassware at the end of the experimentation, so that it is ready for the next laboratory work.
- Execute the assigned pre-lab exercises (if any). These generally cover any observations, calculations, or important points that need to be noted down.
- Try to make a brief outline of the experiment including reagents and materials required, observations to be made, calculations in the observation notebook, which facilitates a quick start.
- The standard operating laboratory procedures should be followed exactly as they are given in the literature. Any modification, if necessary, shall be tested through trials with replications, before getting it published.
- The data format should be ready before proceeding for the experimentation, which not only saves time, but also helps with speedy calculation and data analysis.

Many students forget to write down their observations. Color changes, endpoints, endothermic or exothermic changes, changes in physical state, boiling point, melting point, freezing point, etc., are crucial observations to be recorded. The data collected should suffice the desired experimental objectives and results. It is necessary to replicate experiments to confirm the data whenever any doubt arises. This is also an opportunity to pinpoint the mistakes made during the experimentation and correct the mistakes at the appropriate step(s).

It is pertinent to note that everyone working in the laboratory must strictly follow the safety dos and don'ts of the laboratory, not only for their own safety but also for fellow colleagues. Do not forget to report any dangerous lab practices/events encountered during any experimentation, which will help others to understand and be cautious.

The scientists/lab advisors/technicians must discuss with the students/experiment-performs the important aspects of the experimentation, so that the students will have a better clarity of the experiments they plan to execute

## 1.5 LABORATORY TIPS FOR STUDENTS USING ORGANIC SUBSTANCES

The use of organic chemicals/reagents in an organic laboratory is crucial, wherein the water can also be the death of a reaction. Students usually commit mistakes at various steps, right from washing their glassware. Therefore, every step to be followed carefully, right from washing and drying of the glassware to be used.

- For quick drying of glassware washed with water, dry it with the help of acetone or make use of desiccators.
- Proper labeling of the reagents is mandatory. Many organic liquids are clear and colorless just like water, as are many aqueous solutions such as acids and bases.
- The liquid organic reagents are to be measured by volume. Most common organic liquids have their densities reported in one or more of the reference books (Mass/volume=density).
- The corrosive liquids of some reagents, especially solvents such as acetone, ethyl acetate, acids, even in dilute form, are capable of dissolving some fabrics, like nylons, rayon. etc. Therefore, do not wear your best clothes to lab and always use lab coats/aprons.
- The physical data on common organic chemicals encountered in most laboratories (boiling point, melting point, and density) shall be obtained from the following sources:
    - The CRC Handbook of Chemistry and Physics
    - Aldrich Chemical Company Catalog
    - The Merck Index
- Cooling water always enters the bottom of the condenser and flows out the top at a gentle pace, so no need to blow off hoses and soak your lab mates. Check if the water is running before heating starts.
- Check the glass surface for its temperature before disassembling distillation and reflux equipment.
- Use of special waste containers is necessary, as many organics do not mix or dissolve in water. Do not pour organics down the sink.
- When you are performing extraction or separation, save both the layers of organic and water, and make sure you know the location of the desired product. If you are not sure, add a few drops of water to both from your wash bottle, which will make a clear distinction.
- Don't forget to stop the distillation process well before the boiling flask goes dry. Residues concentrated to dryness or near dryness during distillation may be unstable, which may lead to an explosion. This is particularly applicable with ethers and some alcohols that can form organic peroxides.
- Use of enough grease on glass joints to prevent "freezing" (remember that excessive usage drips from the joints) except for Teflon stopcocks, which are never greased. Glass joints that come in contact with the strong bases such as KOH, NaOH, etc., definitely need to be greased. Glass surfaces may get permanent sealed if not cleaned after use.

- Even if you need to leave the lab for a few minutes for any reason (washroom, coffee, etc.) ensure your colleagues/lab technician/lab assistant watch your reaction. Never leave a reaction unattended.

## 1.6 SAFETY AND QUALITY CONTRACT

Any experimenter, who wants to work in the laboratory, should sign a contract, which shall include the following:

- They will always follow instructions and guidelines given by technical staff.
- They will make the appropriate records to fulfill the demands of the log-book, including any quality control parameters mentioned in the log-book.
- They will clean fume hoods, lab space, glassware, and equipment at the end of every lab session.
- They will identify and use the correct method of disposal for different types of waste and make sure that the waste receptacles are properly labeled.
- They will strictly follow good laboratory practices, including returning materials to their correct locations, and proper care and handling of equipment.
- They will always use a lab coat/apron inside the laboratory.
- They will use protective glasses, when they are working with acids, bases, and other corrosive chemicals.
- They will plan and finish the work in the laboratory within normal working hours 9:00–17:30 of working days (Monday–Friday/Saturday).
- They will have a valid pass signed by the supervisor and laboratory in charge if they have to work in the laboratory outside normal working hours and will follow the instructions given on the pass.
- They will report any injury or a mishap immediately to the instructor/laboratory in charge/the person responsible for the room.
- They will ensure closing of taps and switching off power points at the end of every work.

## 1.7 LABORATORY SAFETY: GENERAL SAFETY RULES

(Oregon State University, Health and Safety Training Manual, Section 2, Chapter 16)

### 1.7.1 General

- Safety takes precedence over all other considerations.
- Never work alone while handling/performing dangerous chemical procedures.
- Ensure that there is someone in the immediate vicinity that you can reach in case of emergency.
- Confirm the location of and how to use eyewash fountains, deluge showers, and fire blankets, in case of emergency.
- Ensure that you understand the hazards involved in a procedure and take all necessary safety precautions before beginning.

- Unsafe facilities, activity equipment, or behavior should be immediately reported to the supervisor/laboratory in charge.
- Food products (lunches, snacks, juices, condiments, etc.) are not to be stored in laboratory refrigerators. Consumption of food and beverages is highly prohibited. Smoking is not permitted in any of the laboratory premises/operation areas.
- Unattended equipment and reactions are major causes of fire, floods, and explosions.
- Anticipate hazards that would result from failure of electrical, water, or gas supply and plan for safety precautions and emergency operations.

### 1.7.2 Personal Protection, Clothing, and Hair

- Ensure that all containers are properly labeled.
- Don't forget to wear approved eye and face protection suitable for the work at hand.
- Safety glasses or goggles are to be worn at all times while working with chemicals at the counter or laboratory hood.
- When working with potentially eruptive substances, use of face protection is highly necessary.
- Custodians, maintenance workers, and visitors must observe safety rules, including eye protection, while in the laboratory.
- Wearing of protective gloves and clothing is mandatory whenever handling corrosive or other hazardous chemicals/reagents.
- Wear closed-toe shoes at all times in the lab.
- The moving parts of mechanical apparatus are guarded to prevent hazardous contact.
- The laboratory area must be kept reasonably neat, clean, and uncluttered.
- Use the fume hood for all operations involving harmful gases or fumes and for flammable or explosive materials. Check the hood to see that it is working properly and subjected to annual/periodical inspection.
- When performing the reactions with large volume of flammable liquids or unstable material, always use a safety shield or barrier to protect against explosion, implosion, and flash fires.
- Remember to inspect glassware for cracks, sharp edges, and contamination before using. Broken or chipped glassware must be repaired and polished or discarded.
- Always remember to use a lubricant (e.g., water, glycerol) whenever attempting to insert the glass tubes into the rubber stoppers or grommets. Protect hands in case tubing breaks.
- Broken glass should be put in impervious containers that are large enough to completely contain the glass. These containers are to be placed into the building trash dumpsters by laboratory personnel.
- Never handle radioactive isotopes without concurrence of the radiation safety officer.

### 1.7.3 Chemical Handling

- Never attempt to work with large quantities of reactants without special precautions.
- Remember to use a safety pail for transporting dangerous or flammable liquids of more than a small quantity (one pint). Prevent the tipping of containers when transporting materials on a cart.
- Do not pour anything back into a reagent bottle.
- Do not pipette the reagents chemicals by mouth or manually. Make use of stripettes/pipette filler/automatic pipettes.
- Be cautious whenever adding anything to a strong acid, caustic, or oxidant. Remember to add slowly.
- Never attempt to add solids (boiling chips, charcoal, etc.) to a hot liquid.
- Never point the mouth of a vessel being heated toward any person, including yourself.
- Guard yourself against infection by skin contact, inhalation of aerosols, and contamination of food and beverages, when working with bio-hazardous materials.
- The known carcinogens, mutagens, and teratogens should not be used or stored in normal laboratory situations. Such substances require extreme precautions, tight security, limited access, and appropriate safety procedures, and should be used in conjunction with the prescribed standard carcinogen safety program.
- Always locate energized electrical equipment or other devices that may emit sparks or flame at least six inches above the floor.
- Do not heat flammable solvents in an open vessel in the presence of sparks or flame. Remember to use only steam, hot water, or a grounded heating mantle for heating flammable liquids.
- The experiments involving ether and other volatile flammable liquids should be considered as fire or explosive hazards and proper care must be taken accordingly.
- Ensure that the natural gas lines in the laboratory are shut off at the line valve rather than at the equipment when not in use.
- It is essential that electrical apparatus is properly grounded. Except for dual-insulated equipment, laboratory electrical apparatus should have a three-conductor cord that connects to a grounded electrical outlet.
- The electrical wiring used for experiments, processes, etc. should be done neatly as per the standard norms, and they must conform to electrical safety code requirements.
- Store the strong oxidants such as nitrates, chlorates, perchlorates, and peroxides in a dry area apart from organic materials.
- Acid digestion must be done in specially designed wash-down laboratory hoods.

### 1.7.4 Chemical Storage

- It is mandatory that all chemical substance containers shall be labelled to identify contents. All flammable liquid containers shall be labeled as "Flammable" or "Ignitable."

- Flammable solvents should be stored in approved, flammable-liquid storage cabinets, or in approved solvent-storage rooms. Not more than 37.8 liters of flammable liquids combined shall be stored in the laboratory outside of approved storage mentioned above.
- Unsealed containers of peroxide-forming compounds should never be stored in the laboratory, as the organic peroxides may detonate by shock, friction, or heat. Certain ethers, unsaturated hydrocarbons, aldehydes, and ketones are the compounds that have dangerous tendencies to form peroxides by reaction with oxygen. As these peroxide-forming compounds have a limited shelf life, they should never be stored for longer than one year.
- Never store caustic liquids above eye level.
- Glass containers of hazardous liquids are not to be stored on the floor, unless they are inside protective containers or pans.
- Inspect the chemicals/reagents periodically and discard old, no-longer-needed substances through the campus hazardous waste disposal program.

### 1.7.5 Pressure and Vacuum Systems

- Never perform experiments that develop high pressure or vacuum unless the consequences of an explosion have been considered and provided for.
- Do not attempt to heat the reactants of any kind in a fully closed system without an approved pressure release system.
- It is advisable not to open a pressurized vessel (autoclave, etc.) until pressure has been fully released.
- The compressed gas cylinders must be kept in an upright position at all times to prevent them from falling. Do not move or store the compressed gas cylinders without the protective caps in place.
- Never attempt to interchange the regulators designed for specific cylinders.
- Flammable gas cylinders must not be stored next to exits or oxygen cylinders.
- Use only approved racks or securing devices for moving of bottled gas cylinders. Do not use the lift truck or hand truck.
- Never use oxygen as a substitute for compressed air. Do not use oil on gauges or regulators for oxidizing gases, as the oxygen under pressure reacts violently with oil or grease.
- Never use compressed gas from a cylinder without a reduction of pressure through a suitable pressure regulator.
- Pressure-adjusting screws on regulators shall always be fully released before the regulator is attached to a cylinder. Always open the valves on cylinders slowly. Do not stand in front of pressure regulator gauge faces when opening cylinder valves.
- Do not strike the valves with tools, or use excessive force in making connections.
- Avoid the mixtures of acetylene and oxygen or air prior to use except at a standard torch.
- Cylinders not provided with fixed hand-wheel valves shall have keys or handles provided on valve stems at all times when cylinders are in use.

- Cylinders should never be dropped, bumped violently, skidded or rolled horizontally. Compressed gas cylinders are high-pressure vessels and should be handled accordingly.
- Do not store cylinders in direct sun, or in boiler or furnace rooms.

### 1.7.6 Container Handling

- Ensure that all containers are properly labeled.
- Do not reuse a food container without first removing the original label completely.
- Chemical transport containers are not to be used for non-compatible chemicals or for food products at any time.
- Never place uncapped vessels of chemicals in a refrigerator, on benches, or in hoods.
- Refrigeration of flammable materials must be done in spark-proof or explosion-proof refrigerators.

### 1.7.7 Chemical Spills and Disposal of Chemical Wastes

- Devise a plan to deal with spills before they occur, and post the plan in the laboratory. Quickly and thoroughly clean up any liquid or solid chemical spill in the laboratory or area of operations. If any uncertainty exists, seek the assistance of supervisor or call Environmental Health and Safety.
- Chemical waste is to be disposed of by approved methods only. Unwanted or no-longer-useful chemicals become chemical waste. Contact Environmental Health and Safety for waste disposal guidelines.
- Prior to disposal, reagent bottles should be thoroughly cleaned of any hazardous materials.

### 1.7.8 Laboratory Precautions and Safety Procedures

- No food or drink is allowed in the laboratory.
- Although the laboratory tables and counters are wiped down before each lab set up, as a result of some laboratory exercises, chemical residues may be present on the tables. Therefore, for laboratory exercises/experiments involving food or drinks, lab assistants and instructors should follow procedures that allow safe consumption.
- Shoes must be worn in the laboratory.
- If you have very long hair, tie it back when you are working in the laboratory.
- Handle chemicals, reagents, and stains carefully and follow all warnings and standard operating procedures. All bottles and containers are to be labeled as to contents and potential hazards.
- For potentially hazardous chemicals, information on the hazards, proper handling, and clean-up is provided on Material Safety Data Sheets (MSDS). These are available in the laboratory. It is highly recommended that you read and understand the contents of MSDS.

- Read all precautions in the laboratory manual and on labels and follow the appropriate directions exactly.
- Always double-check the name of the reagent to be used.
- Reagent bottles should remain stoppered, except when you are actually pouring solutions out of them. Always replace the stopper or lid of stock solutions or stains. Ensure to put them on the container they came from.
- Take only as much as you need and never return leftover solutions to a reagent bottle. Discard leftovers in the proper container.
- Do not use your thumb as a stopper. Swirl gently or put a piece of parafilm over the opening to mix solutions.
- Label all test tubes, flasks, and other containers with contents.
- Do not pour reagents and chemicals down the sink. Dispose of them only in designated containers.
- Do not discard solids or plant materials down the sinks. They will clog up the sink and arrest the easy flow passage. Always use the specified containers for the disposal of such waste.
- Be cautious whenever using razor blades to prepare lab materials. Put the used blades into the designated container from which you obtained them. Do not leave them loose on the desks or counters. Never put razor blades in the wastepaper basket.
- If glassware accidentally becomes broken, carefully clean it up with a broom and dustpan. Dispose of the broken glass pieces in containers labeled for broken glass only.
- When the hot plates are being used for any experimentation, unplug them before you leave the lab. If you notice that the cord on a hot plate or microscope has become frayed and wires are showing, report this immediately to the instructor/laboratory in charge.
- If you are in doubt about directions for an experiment or activity or about use or disposal of materials, ask first before acting.
- Your instructor/laboratory in charge will review with you the location and, where applicable, use of the safety equipment in the laboratory including MSDS files, emergency phone, first aid kit, fire extinguisher, and eyewash.
- When preparing the solutions for your experiments or stains for experiments or observations, put the containers or bottles back in their designated place after you have used them.
- Greenhouse materials, for example, are often sprayed with hazardous materials. Therefore, do not take away the specimens, unless directed to do so.
- Do not waste the paper towels or any other lab materials.
- You are expected to clean up after yourself:
    - Take labels from test tubes and/or beakers.
    - Wash all the glassware and put it in the right place.
    - Place all the dirty slides and cover slips in the designated containers.
    - Ensure tables are clean when you leave the lab.
    - Throw away any trash you generate.
    - Wipe up water, other liquids, soil, and plant material.

- Be sure to discard extra plant material in the designated containers.
- Return the prepared slides in the proper slide tray.
- Take the slide off the stage of your microscope, put the scanning (3X) lens in place, and return your microscope to its proper place in the microscope cabinet.

### 1.7.9 Laboratory Setup

Some important features should include:

- A dedicated vented space, separated from the office areas by an interior door.
- An exterior door or window or alternative interior exit opposite the main interior entrance that can be opened and used for emergency egress.

Organize the workspace, including bench and instrument location, to maximize efficiency and minimize the risk.

### 1.7.10 Laboratory Workbenches

Ideally, lab workbenches should be at the following heights:

- Precision work (e.g., instrument repair or maintenance): Workbench height should be above elbow height (95–110cm above the floor).
- Light work (e.g., taking pH or EC readings): Workbench height should be just below elbow height (85–95 inches above the floor).
- Heavy work (e.g., grinding or processing soil): Workbench should be 10–15cm below elbow height (70–90cm above the floor).

### 1.7.11 Preventive Measures for Laboratory Bench Work

#### 1.7.11.1 General

Observing the following can reduce the risk of injury:

- Always assume the proper sitting or standing posture.
- When sitting, use only an adjustable stool or a chair with a built-in footrest to ensure lower back, thigh, and foot support, or use an external footrest.
- If leg clearance is not available, the workbench must not be used for seated work.
- When standing for extended periods of time, use an anti-fatigue mat and a foot-rest for propping one foot up at a time to reduce joint strain and muscle fatigue.
- Take frequent short breaks to interrupt repetition, awkward body posture, and static muscle work.

#### 1.7.11.2 Pipetting

Pipetting involves thumb force, repetitive motion, and awkward postures for the wrists, arms, and shoulders. Observing the following can reduce the risk of injury:

- Perform your work only at appropriate heights.
- When sitting, ensure lower back and thigh support by using an adjustable stool or chair with a built-in footrest, or use an external footrest.
- Work with wrists in neutral positions (straight).
- Adjust the height and position of sample holders, solution container, and waste receptacle to prevent twisting and bending of your wrist and neck. Keep items close to avoid reaching.
- Reduce shoulder strain by arranging work to keep elbows and arms close to the body.
- Use the short pipettes and shorter waste receptacles for used pipette tips in order to reduce reaching.
- Use electronic pipettes for highly repetitive pipetting tasks or for larger workloads to reduce or eliminate contact pressure on thumb.
- Use ergonomic pipettes when possible.
- Alternate continuous pipetting with other tasks, or take short rest breaks every 20 minutes.

#### 1.7.11.3 Microscopy

Operating a microscope for long hours puts strain on the neck, shoulders, eyes, lower back, arms, and wrists. Observing the following can reduce the risk of injury:

- Make sure leg and knee clearance under workbench is adequate.
- Always assume proper sitting position. Ensure proper lower back and thigh support. Maintain natural curves of spine.
- Ensure that feet are flat on floor or supported by a footrest.
- Adjust the microscope's eyepiece height to allow for natural posture of the head and neck. The vertical position of the eyepiece should be a little high, so that your head is upright.
- Position the microscope as close to you as possible to ensure upright head position.
- Do not work with your elbows winged. Keep your elbows close to your sides.
- Make sure to work with wrists in a neutral (straight) position. Avoid forearm and wrist contact pressure. Pad sharp edges with foam, or pad wrists and forearms to reduce pressure.
- Use a video display terminal when appropriate to view sample, reducing eye and neck strain.
- Make sure that scopes remain clean all the time and that lighting is of proper intensity.
- Take mini-breaks often to allow muscles to relax and to stretch.

#### 1.7.11.4 Fume Hoods

If your laboratory does not have a fume hood, in instances where work requires a fume hood, a local college or university can often serve as a useful resource by providing access to one. Working in fume hoods requires personnel to assume a variety of awkward postures because of limited work access. These restrict arm movement and therefore significantly increase the amount of stress on joints of the upper limbs, neck, and back. Observing the following can reduce the risk of injury:

- Prevent over-extension by placing materials as close as possible. Perform your work at least 15cm back into the hood to maintain optimal airflow containment for material and personal protection.
- Always assume a proper posture. Use only an adjustable chair or stool, and use a footrest if your feet do not rest firmly on the floor.
- Avoid contact pressure of the forearm and wrists on sharp edges. Apply foam padding to the front sharp edge of the fume hood to reduce pressure.
- Use a turntable to store equipment near the worker. This reduces excessive reaching and twisting, which place an increased load on the lower part of the back.
- When standing for extended periods of time, use anti-fatigue mats and a footrest for propping one foot up at a time to reduce joint strain and muscle fatigue.
- Take short breaks to alter repetitive forearm and wrist motion and to relieve joint pressure and contact pressure caused by sharp edges.
- Reduce eyestrain and awkward posture by keeping the viewing window of the hood clean and your line of sight unobstructed.

### 1.7.12 Recommended Safety and Emergency Equipment for the Laboratory

#### 1.7.12.1 Personal Protective Equipment (PPE)

- Chemical splash goggles
- Face shields
- Lab coat
- Lab apron
- Gloves (selected on the basis of the material being handled and the particular hazard involved)

#### 1.7.12.2 Safety and Emergency Equipment

- Hands-free eyewash stations (not eyewash bottles)
- Deluge safety showers
- Safety shields with a heavy base
- Fire extinguishers (dry chemical and carbon dioxide extinguishers)
- Sand bucket
- Fire blankets
- Emergency lights
- Emergency signs and placards
- Fire-detection or alarm system with pull stations
- First-aid kits (special note: if hydrofluoric acid is used, be sure to have a calcium gluconate kit on hand)

- Spill-control kit (absorbent and neutralizing agents)
- Chemical-storage cabinets (preferably with an explosion-proof ventilation system)
- Gallon-size carrying buckets for chemical bottles
- Laboratory chemical hood (18–30m per minute capture velocity, vented outside)—if necessary, use a fume hood at a local college or university for procedures requiring one
- Container for broken glass and other sharp objects
- Safety Data Sheets for all chemicals in the laboratory
- Emergency Action Plan

# 2 Basic Principles of Analytical Techniques and Instrumental Methods

## 2.1 INTRODUCTION

Instrumental methods of chemical analysis are the quantitative methods of analysis that use, as principal measuring devices, instruments other than those employed in conventional gravimetric and volumetric determinations.

The main basis for this method of analysis is the physical property of any particular element or compound being analyzed. These properties include:

- Mass/volume
- Specific gravity, surface tension, viscosity, velocity of sound
- Properties involving interaction with radiant energy—absorption, scattering, emission
- Electric properties such as half-cell potential, current voltage characteristics, electrical conductivity (EC), dielectric constant, magnetic susceptibility
- Thermal properties—thermal conductivity, heat of reactions, transition temperature
- Nuclear properties—radio activity, isotopic mass, etc.

## 2.2 ADVANTAGES OF INSTRUMENTAL METHODS

- Time saving
- Large number of samples can be handled at a time
- Variety of methods available, hence we can choose alternate methods
- Instruments have higher sensitivity and there is minimum personal error
- Wider applicability—solids as well as liquids can be determined

DOI: 10.1201/9781003430100-2

## 2.3 CLASSIFICATION OF INSTRUMENTAL METHODS

| | |
|---|---|
| Electrometric methods | Electrolysis, conductometry, potentiometry, polarography, amperometry, coulometry |
| Optical methods | Colorimetry, turbidimetry, spectrophotometry, flame photometry, spectrography, polarimetry, X ray emission, refrectometry |
| Thermal methods | Differential thermal analysis |
| Gas, volumetric methods | Adsorption, combustion |

## 2.4 BASIC PRINCIPLES OF INSTRUMENTAL METHODS

### 2.4.1 Conductometry

Solutions of electrolytes conduct an electric current by the migration of positively charged species towards the cathode and negatively charged anions towards the anode under the influence of an electric field. Solution of strong acids, strong bases, and most salts are good conductors of electric current and obeys Ohm's law.

The resistance (R) is expressed in Ohms and the conductance of the solution is the reciprocal of its resistance and is expressed as 1/Ohms or mhos. Specific conductance is the conductance given by the solution between electrodes of $1 cm^2$ and 1cm apart.

Equivalent conductance is the conductance of a solution containing 1g equivalent weight of dissolved electrolyte between electrodes 1cm apart. As solutions become dilute, their equivalent conductance increases.

$$\text{Equivalent conductance} = (1000 \times \text{specific conductance}) / \text{Normality}$$

According to Ohm's law for the same type of material, the resistance is proportional to the length of the conductor and inversely to the cross section or diameter. The resistance offered decreases with electrolyte concentration in the solution and hence, the intensity of current flowing through the solution is directly proportional to the electrolyte concentration.

In conductivity measurements, the intensity of current flowing between two electrodes of known and constant surface area through the electrolyte solution is measured, keeping the distances between the electrodes constant.

A conductometric titration involves the measurement of the conductance of the sample after successive addition of the reagent. The end point is determined from a plot of either the conductance or the specific conductance as a function of the volume of added titrant.

### 2.4.2 Potentiometry

Potentiometric titrations make practical use of the relationship between the concentrations of the components of the cell and the EMF of a voltaic cell. That is, near the end point, there is a sudden change in the EMF or pH.

Potentiometric methods are highly useful in titrations involving turbid of highly colored solutions where the use of chemical indicators is either unsatisfactory or impossible.

In potentiometric titrations, the pH values of solutions are experimentally determined during the progress of titration. By plotting these values against corresponding volumes of titrating solutions added, curves with characteristic patterns are obtained. From these curves, the equivalence points can be obtained by finding the midpoint of the inflection of the curve or by finding the point of maximum slope of the curve.

Potentiometric measurement is based on the Helmholtz electrode potential and Nernst equation as follows:

$$E = (RT / nF) \times 2.303 \text{ pH}$$

$$E = 0.058 \times \text{pH}$$

$$\therefore \text{pH} = E, \text{ since } 0.058 \text{ is negligible}$$

The potential difference is measured electronically by immersing a glass electrode with standard KCl-Calomel reference electrode. The potential measured in millivolts is recalibrated into pH units using the following formula of $\text{pH} = (E - 0.246) / 0.058$.

### 2.4.3 Colorimetry

Colorimetry is quantitative analysis in which the quantity of a colored constituent is determined by measuring the relative amount of absorption of light passing through a solution of the constituent.

Comparison is made directly or indirectly, by either visual or electrical means, between the intensity of light passing through the solution and the intensity of light of the same wavelength passing through a solution (or series of solutions) of the same substance at known concentrations.

Coloring of the substances is of three types as follows:

1. The constituent itself may be colored; e.g., determination of $KMnO_4$ ion.
2. The constituent may be made to react with a substance to form a colored compound; e.g., estimation of P in soils.
3. The constituent may be made to displace an equilibrium existing between two colored compounds and thus give a definite, determinable intermediate shade of color to the solution; e.g., determination of pH of a solution from the color produced by addition of an indicator.

Colorimetric methods are generally limited to determinations at relatively low concentration (< 2%). To understand colorimetry, it is better to have a basic knowledge of light and its properties.

Light passes in the form of electromagnetic radiation. It is a form of energy transmitted at an enormous velocity. It is transmitted as waves having electric and magnetic components. The visible (light) range occupies only a small region in the spectrum of electromagnetic radiations.

Wavelength is the distance between two maxima of an electromagnetic wave. It is expressed as meters, centimeters, millimeters, micron (μ) (1 in million), millimicron (mμ), nanometer (nm), and Angstrom (A°) units.

$$1A° = 10^{-8}cm = 10^{-7}mm = 10^{-10}m = 0.1nm$$

$$1nm = 1m\mu = 10^{-9}m = 10^{-7}cm = 10^{-6}mm = 10A°$$

The major components of electromagnetic spectrum are as follows:

| | |
|---|---|
| Gamma ray region | 0.02–1A° (shortest waves emitted by atomic nuclei) |
| X ray region | 1–10A° (emitted due to movement of electron close to the nuclei of heavy atoms) |
| Ultraviolet region | 1–180nm |
| • Vacuum ultraviolet | 180–400nm |
| • Ultraviolet | |
| Visible region | (400–750nm) |
| • Violet | 400–435nm |
| • Blue | 435–480nm |
| • Green blue | 480–490nm |
| • Blue green | 490–500nm |
| • Green | 500–560nm |
| • Yellow green | 560–580nm |
| • Yellow | 580–595nm |
| • Orange | 595–610nm |
| • Red | 610–750nm |
| Infrared region | 0.7–2.5μ |
| • Near infrared | 2.5–15.0μ |
| • Infrared | 15–200μ |
| • Far infrared | |
| Microwave region | 0.1mm–1cm |
| Radio frequency region | 1 cm–10m |

The energy of electromagnetic radiations is inversely related to the wavelength. The energy of these electromagnetic radiations decreases from gamma to radio frequency waves.

Colorimetry is the determination of the concentration of a substance by measurement of the relative absorption of light or transmittance with respect to a known concentration of the substance. It is mainly based on Lambert's and Beer's laws.

According to Lambert's law, when a monochromatic light passes through an absorbing medium, its intensity decreases exponentially as the length/thickness of the medium increases.

According to Beer's law, the intensity of ray of monochromatic light passing through an absorbing medium decreases exponentially as the concentration of the absorbing medium increases.

Mathematically,

Lambert's law $T = (I / I_0) = 10^{-KL} = \log_{10} (I_0 / I) = K^1L$

Beer's law $T = (I / I_0) = 10^{-KC} = \log_{10} (I_0 / I) = K^1C$

Lambert's and Beer's laws $T = (I / I_0) = 10^{-KCL} = \log_{10} (I_0 / I) = KCL$

Where,

| | | |
|---|---|---|
| $I_0$ | = | Intensity of incident light |
| I | = | Intensity of transmitted light (light leaving the solution) |
| K | = | Constant |
| L | = | Length of the absorbing medium |
| C | = | Concentration of the solution |
| T | = | Transmittance |

From the above,

$$\text{Absorbance (A)} = \text{Log}_{10}(1/T)$$

Where, T = Transmittance

The essential components of a colorimeter are as follows:

- A source of radiant energy
- Some type of filter for isolation of a band of radiant energy
- An optical system for producing a parallel beam of filtered light for passage through an absorbing cell
- A detector
- Associated readout meters

In the colorimeter, the light converted to monochromatic by passing through a filter is passed through water/blank solution and the instrument is adjusted to 100% transmittance or zero optical density/absorbance. Then the solution containing the color producing substance (chromogen) at various concentrations is placed in turn in the light path. Absorbance can also be calculated from the observed transmittance by using the following formula.

$$\text{Absorbance} = 2 - \log_{10} (\text{Transmittance})$$

From the optical density or transmittance is measured from the scale, a standard graph is prepared. Then the solution with the unknown concentration is placed and the optical density or transmittance is read from the scale. By referring to the standard graph, the actual concentration is determined.

### 2.4.4 Spectroscopy

Spectroscopy is the study of the spectral properties. The spectrum of light into the colors of which it is made is known as dispersion and the band produced by the dispersion is called visible spectrum.

Generally, prisms or gratings are used to split the white light into seven colors as we see in rainbows. A spectrophotometer is used to measure the intensity of specific spectral line of the light.

Instruments used to measure the intensity of spectra are as follows:

| | |
|---|---|
| Colorimeter | • Natural or artificial white light is generally used as a light source.<br>• It mainly uses a filter to isolate the wavelength.<br>• A photocell or photo tube is used as a detector.<br>• It is used to measure absorption in the visible range. |
| Spectrophotometer | • Light of definite wavelength, 1–10A° band width, extending to the ultraviolet region of the spectrum constitutes the source of light.<br>• Monochromation is by prism or grating. Hence, the resolution of wavelength is high.<br>• Photo tube or photo multiplier scatters to a wide range of wavelength during estimation.<br>• It is used to measure the intensity of specific spectral line of the light. |
| Photometer | • Used to measure the energy of transmitted light.<br>• It is a via media between colorimeter and spectrophotometer. |

Spectroscopy is of three types as follows:

1. Flame emission spectroscopy (FES)/flame photometry
2. Atomic absorption spectrophotometry (AAS)
3. Emission spectroscopy (ES)

### 2.4.5 Flame Photometry/Flame Emission Spectroscopy

Flame photometry is the study of the photon (light) energy emitted when a metal is introduced into the flame. It involves the measurement of the intensity of radiation emitted by the atoms that are excited by the high temperature of the flame. It is an important tool for the determination of alkali and alkaline earth metals for which the chemical methods are time-consuming.

The elements present in the sample, when excited by the flame, emit photons of characteristic wavelength and the intensity of which is related to the concentration of the element concerned.

$$\text{Intensity of radiation } \alpha \text{No of atoms in the sample}$$

Fraction of atoms excited depends critically on the flame temperature. Hence, temperature control of the flame is very important in flame photometry. The important parts of flame photometer are as follows:

- Pressure regulator and flow meters for the fuel gases
- Atomizer
- Burner
- Optical system
- Photosensitive detector
- Instrument for recording the output of the detector

For flame photometry, the sample to be detected is prepared in solution form and sprayed under controlled conditions into a flame. The radiation from the flames enters a dispersing device to isolate the desired region of the spectrum. A photo tube and some sort of meter or electronic amplifier measures the intensity of isolated radiation.

Before initiating the experiment, the flame photometer needs to be calibrated (preparation of a standard graph by employing different concentrations of a solution containing the element to be determined). Flame photometers are generally employed for the estimation of sodium (Na) and potassium (K).

Radiation in the visible and ultraviolet regions occurs when the atoms or molecules are excited by the absorption of energy. Each element has its own characteristic atomic spectrum, which is very important in flame photometry.

The atomic spectrum of an element results when sufficient energy is supplied to volatilize the element and cause the electrons in the atoms or ions to move to higher energy status. The species is then said to be excited. The duration in the excited stage is very short ($10^{-8}$ seconds).

The electrons tend to return spontaneously to their low energy levels frequently in a series of discrete steps. Each of these steps involves the loss of definite amount of energy in the form of electromagnetic radiation. The wavelength of this is determined by the magnitude of the energy difference between levels. As such, there are many possible excited states and multiplicity paths by which return to ground state can occur. Hence, many radiations of different wavelengths are produced by a given element. The collection of radiations for a given species, called emission spectrum is characteristic for that species alone.

Several methods are available for exciting the elements for analysis. One of the methods is heating the sample in a flame and cause excitation. In a flame photometer, emissions from excited atoms constitute the radiation source.

### 2.4.6 Atomic Absorption Spectrophotometry (AAS)

This method is especially useful for the determination of elements that are vaporized but not appreciably excited in oxyacetylene flame. The element of interest in the analysis is made the source of radiant energy and is subjected to electrical excitation causing to emit radiation of its characteristic frequencies. When this beam of light is passed through the vaporized sample interposed between the source and the entrance slit of a monochromator, the atoms of the sample absorb this light and thus decrease the radiant power of the beam.

The radiant power of the spectral line is measured with a photomultiplier tube and amplifier, first without and then with sample on the optical beam. The decrease in radiant power is proportional to the number of atoms of the element of interest (e.g., estimation of micronutrients).

### 2.4.7 Chromatography

Chromatography is a process in which the species of interest are separated by differential migration in a porous medium, the migration being caused by the flow of some mobile phase. Thus, the system consists of two phases, one mobile/moving phase and another stationary phase, both being immersible with each other. There are different types of chromatography as follows:

| | |
|---|---|
| Adsorption chromatography (AC) | High precision liquid chromatography (HPLC) |
| Thin layer chromatography (TLC) | Partition chromatography |
| Gas chromatography (GC) | Molecular sieve and gel permeation chromatography |
| Gas–liquid chromatography (GLC) | Recycling column chromatography |
| Gas–solid chromatography (GSC) | Ion exchange chromatography |

# 3 Analytical Chemistry
## *Basic Concepts*

### 3.1 ANALYTICAL CHEMISTRY

Analytical chemistry is a branch of chemistry dealing with the determination of constituents of a substance or mixture of substances.

### 3.2 QUALITATIVE ANALYSIS

Qualitative analysis deals with the identification of the constituents of a substance, mixture of substances, or solutions and the way in which a component element or a group of elements are constituted.

### 3.3 QUANTITATIVE ANALYSIS

Quantitative analysis is concerned with the determination of quantitative contents of an individual element or group of elements or compound present in the substances. This branch of analytical chemistry is of enormous scope and importance in science, industry, and agriculture. The quantitative analysis can be divided into several categories based on the different methods employed for their quantitative measurements such as:

- Volumetric analysis or titrimetric analysis
- Gravimetric analysis
- Electro gravimetric analysis
- Electro volumetric (potentiometric/conductometric) analysis
- Polorographic analysis
- Colorimetric analysis
- Nephelometric or turbidometric analysis
- Chromatographic analysis
- Radiometric analysis

DOI: 10.1201/9781003430100-3

## 3.4 BASIC PRINCIPLES IN ANALYTICAL CHEMISTRY

A general conception of important fundamentals is essential in analytical chemistry. A number of important principles and definitions are involved in quantitative analysis.

### 3.4.1 Mass and Weight

The mass of a substance is a definite property that can be used as a measure of quantity. The weight of the substance is the force that results from the interaction of the gravitational field on the substance.

Mass can be determined with the help of balance whereas weight should actually be used to express force.

### 3.4.2 Atom and Atomic Weight

An atom is the smallest particle of an element that retains its property and takes part in chemical reaction. Atomic weight is the average weight of an element relative to weight of carbon atom as 12. The unit of atomic weight of an element is amu atom$^{-1}$ (atomic mass unit per atom). The atomic weight expressed in grams is known as gram atom (e.g., 16g of oxygen means one gram atom of oxygen).

### 3.4.3 Molecule and Molecular Weight

A molecule is the smallest particle of substance (element/compound) that retains the property of that substance and exists in free state.

The molecular weight of a substance is defined as the weight of one molecule of the substance relative to the weight of carbon atom as 12. The molecular weight is sum of the atomic weights of the elements present in one molecule of the substance. When molecular weight is expressed in grams, it is termed as gram molecule or mole (e.g., 32g of oxygen means one gram molecule or one mole of oxygen).

### 3.4.4 Avogadro's Number

The actual number of molecules present in one mole of substance is known as Avogadro's number and has the value of $6.024 \times 10^{23}$. Thus, one mole of every substance contains $6.024 \times 10^{23}$ molecules. This quantity also expresses the number of atoms in one gram atom of an element and hence 1 gram atom of every element contains $6.024 \times 10^{23}$ atoms.

### 3.4.5 Equivalent Weight

The equivalent weight of a substance is defined as the weight of an element or compound, which combines with or displaces from combination, eight parts by weight of oxygen or 1.008 parts by weight of hydrogen, or 35.5 parts by weight of chlorine. The equivalent weight expressed in grams is termed as gram equivalent weight.

### 3.4.6 Standard Solutions

The solution of accurately known strength or concentration is called as a standard solution. It contains a definite number of gram equivalent or gram mole per liter of the solution.

### 3.4.7 Strength or Concentration of a Solution

Strength of a solution refers to the weight of a solute dissolved in one liter or per unit volume of the solution. It can be expressed in many ways as follows.

#### 3.4.7.1 Normal Solution or Normality

A normal solution is one that contains one gram equivalent (equivalent weight in grams) of the active reagent, dissolved in one liter of solution.

Normality is the number of gram equivalent of substance, dissolved in one liter of solution. If the number of gram equivalent is one, it is expressed as 1*N*.

Normality = Number of gram equivalent of solute / Number of liters per solution

#### 3.4.7.2 Molar Solution or Molarity

A molar solution is one which contains a gram molecular weight of solute dissolved in one liter of solution. It is denoted by **M**, whereas molarity is the number of gram molecule of substance dissolved in one liter of solution.

#### 3.4.7.3 Molal Solution

A molal solution is one that contains a number of gram molecules of the solute dissolved in 1000g of solvent.

#### 3.4.7.4 Mole Fraction

A mole fraction is defined as the ratio of the number of moles of solute to the number of moles of solution.

#### 3.4.7.5 Formal Solution

A formal solution is one which contains a formula weight of a solute in a liter of solution and is denoted by **F**. In most of the cases, formula weight and molecular weight are identical but sometimes the true molecular weight of a compound is a multiple of the weight expressed by its formula as ordinarily written in a chemical reaction.

#### 3.4.7.6 Percentage Composition by Weight

The concentration is expressed in terms of grams of solute per 100g of solution. For instance, a 10 percent KI solution is prepared by dissolving 10g of the salt in 90g of water.

#### 3.4.7.7 Strength or Percentage Strength

Strength of a solution means the grams of solute or substance dissolved in one liter of solution and usually expressed as g/liter. Percentage strength means the grams of solute per 100mL of solution.

#### 3.4.7.8 Parts per Million (ppm)

The concentration is expressed in terms of grams of solute per milliliters of solution or milligrams of solute per liter of solution. Thus, a solution containing 10mg of solute per liter of solution or 10 microgram of solute per milliliter of solution is 10ppm solution.

#### 3.4.7.9 Milli Equivalent per Liter

A solution containing milligram equivalent of substance in a liter of solution is called Meq/liter.

### 3.4.8 Titration

Titration is defined as the process of determining the volume of a substance required to just complete the reaction with a known amount of other substances. The quantitative analysis carried out by titration is known as "titrimetric analysis."

### 3.4.9 Titrant

The solution of accurately known strength used in the titration is called "titrant."

### 3.4.10 Titrate

The substance (in solution) to be determined by titration is called "titrate."

### 3.4.11 Equivalence Point or End Point

The point (in burette reading) at which the reaction is just complete in a titration is called the equivalence point or end point or stoichiometric end point. At this volume of titrant, the reaction is completed quantitatively with a known volume of titrate.

### 3.4.12 Indicators

A substance that indicates the end point or completion of reaction is called an indicator. It is an auxiliary reagent that helps in the visual determination of the completion of titration. Indicators used commonly in volumetric analysis are of three types as follows.

#### 3.4.12.1 Internal Indicators

The indicators added into the solution where reaction occurs are called internal indicators, which give a clear visual change (either a color change or the formation of a turbidly) in the solution being titrated. Examples of internal indicators are:

- Methyl orange
- Phenolphthalein
- Methyl red
- Methylene blue
- Diphenylamine (DPA)

Internal indicators are divided into three groups according to their use in different types of reactions.

##### *3.4.12.1.1 Indicators Used in Acid–Alkali Neutralization Reactions/Hydrogen Ion/Acid–Base Indicators*

The hydrogen ion (acids-base) indicators are organic dyes (weak organic acids or bases) that change color within limits with variation in pH value of the solution to which it is added.

- Methyl orange
- Phenolphthalein
- Methyl red
- Bromothymol blue
- Phenol red
- Thymolphthalein

##### *3.4.12.1.2 Indicators Used in Precipitation Reactions*

- Potassium Chromate
- Ferric ammonium alum

##### *3.4.12.1.3 Indicators Used in Oxidation–Reduction Reactions*

- Diphenylamine (DPA)
- Methylene blue
- Eriochrome black-T (EBT)

#### 3.4.12.2 External Indicators

There are certain indicators that are not added to the reacting medium or kept away from the solution where reaction takes place, these are called external indicators (e.g., potassium ferricyanide).

#### 3.4.12.3 Self Indicator/Auto Indicator

When one of the reacting substances itself acts as indicator by visual color change, it is called a self or auto indicator (e.g., potassium permanganate).

# 4 Laboratory Vessels and Their Uses

## 4.1 BEAKERS

- Beakers are the one of most important vessels used in the laboratory. They are made of heat-resistant glass.
- They are available in various capacities ranging from 10mL to 3000mL.
- For convenience of pouring solution, a spout is provided on the beaker.
- They are used to perform titrations and for boiling solutions, etc.
- Beakers are constructed in tall and low forms.

## 4.2 PIPETTES

- Pipettes are used to deliver a definite volume precisely. They are of two kinds:
  - Transfer or volumetric pipettes, which deliver a definite volume of liquid under specified conditions.
  - Graduated or measuring pipettes, in which the stems are graduated, are employed to deliver various small volumes at discretion.
- The transfer pipettes consist of a jet, a bulb, and a tube bearing a mark.
- Sometimes, the pipettes are provided with a safety bulb above the mark.
- Transfer pipettes are constructed with capacities ranging from 1mL to 100mL (1, 2, 5, 10, 15, 20, 25, 50, and 100mL) and are the most frequently used kind.
- Graduated pipettes are constructed with capacities ranging from 1mL to 10mL (1, 2, 5, and 10mL). These are less precise than transfer pipettes.
- Micropipettes are also available and they are constructed with capacities ranging from 10 to 100 lamda (1 lamda = 0.001mL).
- All these pipettes are calibrated to contain to deliver a definite volume of the solution in the given temperature, usually 20°C to 30°C.
- In using such pipettes, they are first rinsed with the liquid and filled by suction to about 1–2mL above the mark and the upper end of the pipette is closed with a dry index finger.
- By gently releasing the finger, the level of the solution may be brought to the point where the level of the meniscus coincides to the graduation mark on the pipette.

DOI: 10.1201/9781003430100-4

- The pipette thus filled will deliver exactly the volume of solution that is indicated on the bulb of the pipette.
- The solution is allowed to run out of the pipette into a clean vessel.
- When the continuous flow has ceased, the jet is held in contact with the sides of the vessel for 15 seconds (draining time).
- At the end of the draining time, the tip of the pipette is removed from the contact.
- The liquid remaining in the jet of the pipette must not be removed either by blowing or by other means.
- Some pipettes are calibrated so that the liquid remaining in the jet can also be blown.
- Rubber bulbs are available for handling corrosive or toxic solutions.

## 4.3 BURETTES

- Burettes are long of uniform bore throughout the graduated length, terminating at the lower end in a glass cock and a jet.
- In cheaper varieties, a stop cock may be replaced by a pinch cock.
- The size varies from 10mL to 100mL (10, 20, 50, and 100 mL).
- The micro burette of 5mL and 10mL capacities are helpful in delivering volumes of 0.02mL.
- Before use, the stop cock of the burette must be lubricated with Vaseline.
- The burette is first rinsed with small quantities of titrant solution.
- The burette is then filled to the mark slightly above the zero mark and the solution is made to run down so that the jet is free of air bubbles.
- Then the burette must be firmly attached to the stand.
- The liquid level is adjusted so that its lower level of meniscus coincides with zero level of the burette by using the stop cock.
- To deliver liquid from burette to vessel, place the fingers of the left hand behind the burette and hold the stop cock between thumb, fore, and middle fingers.
- During the delivery of the liquid, the flask may be gently rotated with the right hand to ensure that the added liquid is well mixed with any existing contents of the flask.

## 4.4 GRADUATED OR MEASURING CYLINDERS

- Measuring cylinders are graduated vessels with a broad base.
- They are used for approximate measurements of liquid and vary in size from 5mL to 1000mL.
- It is advisable to use the smallest cylinder that will hold the volume being measured.

## 4.5 VOLUMETRIC FLASKS

- Volumetric flasks are flat-bottomed sphere vessels with long narrow necks.
- A thin line marked around the neck indicates the volume that it holds.

- These are available in various sizes ranging from 5mL to 2000mL.
- A ground glass stopper is also provided with the flask.
- Volumetric flasks are used to prepare standard solutions.

## 4.6 CONICAL FLASKS

- Conical flasks are available in various capacities ranging from 50mL to 1000mL.
- They are very useful in performing titrations where the reactions are instantaneous.
- The narrow neck of the flask prevents the entry of external agencies like $O_2$, $CO_2$, etc., which sometimes may interfere with the analysis.

## 4.7 FUNNELS

- Funnels are helpful for filtration and for transfer of liquids to narrow necked vessels.
- For filtration, the funnel should enclose an angle of 60°.
- The most useful size for titration is those with diameters of 5.75cm and 9.00cm.
- The stem should have an internal diameter of 4cm and should not be >15cm long.
- For filling burettes and for transferring solids to volumetric flask, a short stemmed wide neck funnel is used.

## 4.8 PORCELAIN CRUCIBLES AND BASINS

- Porcelain crucibles and basins are employed for operations in which hot liquids are to remain in contact with vessels for prolonged period.
- They are usually more resistant to alkaline solutions than glass.
- They are also used for evaporating and concentrating the solutions.

## 4.9 GLASS WASH BOTTLES AND POLYTHENE SQUEEZE BOTTLES

- Glass wash bottles and polythene squeeze bottles are useful to keep and deliver distilled water for various operations during analytical work.
- They are necessary apparatus used to wash precipitate and rinse off the reagents from glassware.

## 4.10 GOOCH CRUCIBLES

- Gooch crucibles are porcelain crucibles with perforated bottoms.

## 4.11 SINTERED GLASS CRUCIBLES

- Sintered glass crucibles are made up of resistant glass and have a porous disc of sintered ground glass fused into the body.
- The filter disc is made of varying porosities as indicated by numbers from 0–5 (coarsest to finest).

## 4.12 BUCKNER FUNNELS

- Buckner funnels are made up of porcelain and fitted with a porous disc.
- One or two good quality filter papers are placed on the porous disc and filtration is done by suction.

## 4.13 FILTER FLASKS

- Filter flasks are conical flasks fitted with a side vent, which can be attached to a vacuum pump to create vacuum in the conical flask during filtration.

## 4.14 SILICA BASINS AND CRUCIBLES

- Silica basins and crucibles are used for ignition of precipitates.
- Advantages—they are resistant to heat and acids except hydrofluoric and phosphoric acid.
- Disadvantages—they are affected by alkali and carbonates and require comparatively more time for heating and cooling.

## 4.15 PLATINUM CRUCIBLES

- Platinum has high melting point (1773°C). Hence, platinum crucibles can be used for ignition of precipitates at high temperatures.

## 4.16 DESICCATORS

- A desiccator is a covered glass container designed for storage of objects in dry atmosphere.
- It is charged with some drying agents or desiccants like anhydrous calcium chloride, silica gel, activated alumina, anhydrous calcium sulfate, and concentrated sulfuric acid.
- Hot crucibles are usually cooled inside a desiccator before weighing.

## 4.17 MISCELLANEOUS TOOLS

- Miscellaneous tools are the supporting wares in the laboratory.
- They include burette stands, pipette stands, filter stands, tripod stands, wire gauze, clay pipe triangles, spatulas, tongs, test tube holders, test tube stands, pestle and mortar, porcelain tiles, etc.

## 4.18 CLEANLINESS OF GLASSWARE

- Cleanliness of the glassware helps to improve the precision and accuracy of analytical results.
- Glassware is to be cleaned to sparking brightness to get rid of the contaminating radicals and other impurities.

- Glassware may be washed initially in tap water to remove the soluble chemicals and dirt.
- To remove the greasy and water insoluble dirt and stains, hot detergent solutions preferably of 2% strength are helpful.
- The organic impurities and the hard-to-remove stains can be eliminated by using a chemical mixture of commercial grade concentrated sulfuric acid with potassium dichromate salt. The chromic acid oxidizes the dirt and makes it easily removable.
- Sparking brightness of glassware can be obtained by rinsing with commercial grade concentrated hydrochloric acid.
- After removal of dirt and stains, the wares should be washed free of detergent or acid particles with tap water, and finally rinsed with a small quantity of distilled water two or three times to remove the unwanted radicals which may come as a contamination from tap water.
- Washed glassware should be stored in a dust-free environment after drying in hot air oven.

# 5 Basic Techniques of Analytical Chemistry

## 5.1 VOLUMETRIC TECHNIQUES OR TITRIMETRIC ANALYSIS

1. Acid base neutralization reactions—acidimetry and alkalimetry
2. Oxidation—reduction reactions—oxidometry (permanganometry) and dichrometry
3. Precipitation reactions—thiocyanometry and complexometry

$$V_1N_1 = V_2N_2 \text{ (Law of Volumetric Analysis)}$$

It is the analysis wherein determining the concentration of unknown solution using the given solution of known concentration.

## 5.2 IMPORTANT PRIMARY STANDARDS

### 5.2.1 Acids

- Potassium hydrogen phthalate
- Benzoic acid
- Constant boiling hydrochloric acid
- Sulfuric acid
- Potassium acid iodate

### 5.2.2 Bases

- Sodium carbonate
- Mercuric oxide
- Borax

### 5.2.3 Oxidizing Agents

- Potassium dichromate
- Potassium bromate
- Potassium iodate

DOI: 10.1201/9781003430100-5

### 5.2.4 Reducing Agents

- Sodium oxalate
- Arsenious oxide
- Pure iron metal
- Potassium ferro cyanide

### 5.2.5 Others

- Sodium chloride
- Potassium chloride

## 5.3 IODOMETRY

Iodometry is the process in which standard solution of iodine is used. By determining its amount, the equivalent quantity of another element/radical is deduced. This process may be used to determine sulfurous acid, sulphites, sulfurated hydrogen, thiosulfate, arsenite and stannous chloride. A 1% starch solution is used as indicator which gives an intense blue color even with a very small amount of iodine.

## 5.4 ARGENTOMETRY (SILVER NITRATE TITRATIONS)

The metallic halides like chlorides, bromides, and iodides can be estimated by titrating them in a solution against standard silver nitrate solution.

## 5.6 GRAVIMETRIC ANALYSIS

In this analysis, the substance to be estimated is separated from the rest of the sample as an insoluble compound or precipitate of known chemical composition and the final measurement is made by weighing. The precipitate is often ignited before weighing to convert into a substance of a definite chemical composition and of adequate chemical stability (e.g., estimation of sulfate).

# 6 Preparation of Primary Standard Solutions

## 6.1 PRINCIPLE

Standard solutions are the solutions of accurately known concentrations. There are two types of standard solutions, primary and secondary.

A known quantity of the substances is dissolved in a volumetric flask and the volume is made up to the mark. Standard solutions prepared in this way are known as primary standard. This method can be used for preparation of standard solutions of substances that satisfy the following conditions.

- The substances must be chemically pure, i.e., it must not contain impurities in amounts that could affect analytical precision (not more than 0.05–0.1%). In other words, the substances must be of known and high degree of purity.
- The substances must be stable on keeping both in the solid state and in the solution, as otherwise its composition would cease to its formula.
- It must be non-hygroscopic and not easily affected by acid fumes, $CO_2$ etc., inside the laboratory.
- The gram equivalent of the substance should be as large as possible, so that the precision in determining the normality of the solution is high.
- The substance must withstand fairly high temperature, so that it can be dried at 100°C.
- The substance must be freely soluble in water.
- The substance must be preferably solid and crystalline compounds.

## 6.2 CHEMICALS THAT CAN BE CHOSEN AS PRIMARY STANDARDS

- Succinic acid
- Oxalic acid
- Potassium hydrogen phthalate
- Benzoic acid
- Sulfamic acid
- Sodium carbonate
- Borax
- Sodium oxalate

DOI: 10.1201/9781003430100-6

- Potassium dichromate
- Arsenic trioxide
- Sodium chloride

## 6.3 PREPARATION OF 0.1*N* $NA_2CO_3$ OF 250ML

| | | |
|---|---|---|
| Molecular weight of $Na_2CO_3$ | = | 106 |
| Valency | = | 2 |
| Equivalent weight of $Na_2CO_3$ | = | Molecular weight / Valency |
| Therefore, one gram equivalent weight of $Na_2CO_3$ | = | 106 / 2 = 53 |

53 g dissolved in 1 liter gives 1*N* solution

∴ 5.3g dissolved in 1 liter gives 0.1*N* solution

∴ 1.325g dissolved in 250mL distilled water gives 0.1*N* solution of $Na_2CO_3$

### 6.3.1 Reagents Required

- Analar grade $Na_2CO_3$

### 6.3.2 Apparatus Required

- 250mL volumetric flask with stopper
- Funnel
- Chemical balance with weight box or electronic balance

### 6.3.3 Procedure

- Weigh 1.325g of sodium carbonate (AR grade, oven dried at 110°C).
- Transfer this into a 250mL volumetric flask using a funnel.
- Wash the funnel, collecting the washings in the flask.
- Remove the funnel and add about 100mL distilled water to the flask.
- Shake the flask thoroughly well so as to dissolve $Na_2CO_{3.}$
- Add if necessary one more installment of 100mL water and again shake and thoroughly mix the contents ensuring complete dissolution of $Na_2CO_{3.}$
- Make up the volume to 250mL and again shake so as to make uniform solution.
- Label the solution and preserve for future use.
- The label must consist of name of the chemical, strength, date of preparation, and name of the person who prepared the solution.

# 7 Preparation of Secondary Standard Solution of an Acid

## 7.1 INTRODUCTION

More often, if the reagent is not a primary standard, a secondary standard solution must be prepared. The chemical is dissolved and made up or diluted to the desired concentration and then it is standardized against a primary standard or against another reliably known secondary standard solution. Such solutions are called secondary standard solutions.

## 7.2 PRINCIPLE

Equal volumes of all acids/bases containing their gram equivalent weights per liter of the solution react completely with each other.

$$Na_2CO_3 + 2\,HCl \longrightarrow 2\,NaCl + H_2O + CO_2$$

$$Na_2CO_3 + 2\,HNO_3 \longrightarrow 2\,NaNO_3 + H_2O + CO_2$$

$$Na_2CO_3 + H_2SO_4 \longrightarrow Na_2SO_4 + H_2O + CO_2$$

In other words, the product of volume ($V_1$) and normality ($N_1$) of one solution (Acid) is equal to the product of volume ($V_2$) and normality ($N_2$) of another solution (Alkali/ Base).

From this it is clear that $V_1N_1 = V_2N_2$ (Law of Volumetric Analysis)

## 7.3 PREPARATION OF 0.1*N* HCL OF 250ML (AN EXAMPLE)

| | | |
|---|---|---|
| Molecular weight of HCl | = | 36.45 |
| Valency | = | 1 |
| Equivalent weight of HCl | = | Molecular weight / Valency |
| Therefore one gram equivalent weight of $Na_2CO_3$ | = | 36.45/1 = 36.45g |
| Specific gravity of concentrated HCl | = | 1.18g/mL |
| Percent purity of concentrated HCl | = | 35% |
| One liter of 1*N* solution of HCl contains | = | 36.45g of HCl |

DOI: 10.1201/9781003430100-7

One liter of 0.1*N* solution of HCl contains = 3.645g of HCl
To calculate the volume of conc. HCl that contains this amount of HCl
By proportion for 35g of HCl, 100g of HCl is required
Therefore for 3.645g of HCl = (100 / 35) × 3.645 = 10.4g of HCl
To convert this weight of HCl to volume basis, it must be divided by the specific gravity of the acid (1.18), We then have
= 10.4 / 1.18 = 8.8mL HCl
So, to prepare 1 liter of solution of 0.1*N* HCl 8.8mL HCl is required
Then for 250mL 0.1*N* HCl = (8.8 / 1000) × 250 = 2.2mL
Therefore, 2.2mL HCl is required to prepare 250mL of 0.1*N* HCl solution

### 7.3.1 Reagents Required

- Concentrated HCL
- 0.1*N* $Na_2CO_3$
- Methyl orange indicator

### 7.3.2 Apparatus Required

- 400mL beaker
- 10mL measuring cylinder
- 50mL burette
- 10mL pipette
- 250mL conical flask
- 250mL measuring cylinder
- Glass rod

### 7.3.3 Procedure

- Measure out 250mL distilled water into a 400mL beaker using a 250mL measuring cylinder.
- Add 2.2mL (approximately 2.5mL) of concentrated HCl through a 10mL measuring cylinder to the beaker containing water.
- Thoroughly mix it using a glass rod. This is approximately a 0.1*N* HCl solution.
- Now, fill up a burette with 0.1*N* $Na_2CO_3$ (primary standard) and mount it in a burette stand.
- Pipette out 10mL of the approximately 0.1*N* solution of HCl into a 250mL conical flask and add a few drops of methyl orange indicator. Note down the initial reading of the burette.
- Titrate the contents of conical flask against 0.1*N* $Na_2CO_3$. End point is the change of color from red to pale yellow.
- Tabulate the burette readings and find out the volume of 0.1*N* $Na_2CO_3$ used in the titration.
- Repeat the titration until getting concordant titer values.
- Now using the relationship of Law of Volumetric Analysis $V_1N_1 = V_2N_2$, find out the normality of the HCl.

- If the $Na_2CO_3$ consumed is exactly 10mL then according to the above relationship, the normality of the prepared HCl solution is 0.1*N*.
- If not, the normality of the prepared HCl solution may be higher or lower than 0.1*N*.

### 7.3.4 Observation and Calculation

| | | |
|---|---|---|
| Volume of $Na_2CO_3$ consumed in the titration ($V_1$) | = | X mL |
| Volume of 0.1*N* HCL taken ($V_2$) | = | 10mL |
| Normality of $Na_2CO_3$ ($N_1$) | = | 0.1*N* |
| Normality of the HCl prepared ($N_2$) | = | Y *(N)* |
| We know that, $\mathbf{V_1N_1 = V_2N_2}$ **(Law of Volumetric Analysis)** | | |
| $X \times 0.1$ | = | $10 \times Y$ |
| $N_2 = Y$ | = | $(X \times 0.1) / 10$ |

- Calculate the volume of water to be added if the normality is more than 0.1*N* or the volume of concentrated HCl to be added if normality is less than 0.1*N*.

**If the prepared solution is more than 0.1*N***

| | | |
|---|---|---|
| Volume of 0.1*N* HCl taken ($V_2$) | = | 10mL |
| Volume of 0.1*N* $Na_2CO_3$ used in the titration ($V_1$) | = | X mL |
| Amount of water to be added for 10mL HCl | = | X – 10mL |
| Therefore, the amount of water to be added for (250-A) mL HCl solution | = | $[(X - 10) / 10] \times (250 - A)$ mL |

Where, A = volume of the prepared HCl solution already used for rinsing the pipette and or for titration

**If the prepared solution is less than 0.1*N***

| | | |
|---|---|---|
| Volume of 0.1*N* HCl taken ($V_2$) | = | 10mL |
| Volume of 0.1*N* $Na_2CO_3$ used in the titration ($V_1$) | = | X mL |
| Normality of $Na_2CO_3$ ($N_1$) | = | 0.1*N* |
| We know that, $\mathbf{V_1N_1 = V_2N_2}$ **(Law of Volumetric Analysis)** | | |
| $N_2 = Y$ | = | $V_1N_1 / V_2$ |

$N_1 - N_2 = 0.1 - Y = z$ (N)

For the present 250mL of 0.1*N* HCl, we have taken 2.5mL conc. HCl

Therefore, for z (N) = $[(2.5\ XY) / 0.1] \times (250 - A) / 250$mL

- Add either the volume of water or the volume of acid as the case may be to the solution prepared in a 400mL beaker and mix it thoroughly by stirring.
- Repeat the titration and calculation processes until the normality of the HCl prepared attains 0.1*N*.

### 7.3.5 Format of Titration Table

0.1*N* $Na_2CO_3$ vs Approximately 0.1*N* HCl

| Sl.No | Aliquot taken in the pipette (mL) | Burette readings (mL) | | Volume of titrant used (mL) | Indicator | End point |
|---|---|---|---|---|---|---|
| | | Initial | Final | | | |
| | | | | | | |
| | | | | | | |
| | | | | | | |

# 8 Preparation of Secondary Standard Solution of a Base

## 8.1 PRINCIPLE

Equal volumes of all acids/bases containing their gram equivalent weights per liter of the solution react completely with each other.

$$2\,KOH + 2\,HCl \longrightarrow 2\,KCl + 2H_2O$$

In other words, the product of volume ($V_1$) and normality ($N_1$) of one solution (acid) is equal to the product of volume ($V_2$) and normality ($N_2$) of another solution (alkali/base).

**$V_1N_1 = V_2N_2$ (Law of Chemical Equivalent/Law of Volumetric Analysis)**

## 8.2 PREPARATION OF 0.1*N* KOH OF 250ML (AN EXAMPLE)

| | | |
|---|---|---|
| Molecular weight of KOH | = | 56 |
| Valency | = | 1 |
| Equivalent weight of KOH | = | 56 / 1 = 56 |
| One gram equivalent weight of KOH | = | 56 g |
| 56g dissolved in 1 liter gives 1*N* of KOH solution | | |
| ∴ 5.6g dissolved in 1 liter gives 0.1*N* solution | | |
| ∴ 1.4g dissolved in 250mL gives 0.1*N* solution | | |

### 8.2.1 Reagents Required

- 0.1*N* KOH
- Concentrated HCL
- 0.1*N* HCL
- Phenolphthalein indicator

DOI: 10.1201/9781003430100-8

### 8.2.2 Apparatus Required

- 400mL beaker
- 10mL measuring cylinder
- 50mL burette
- 10mL pipette
- 250mL conical flask
- 250mL measuring cylinder
- Glass rod

### 8.2.3 Procedure

- Weigh 1.4g of KOH and transfer this to a 400mL beaker and add 100mL distilled water and dissolve it by stirring with a glass rod.
- Add the remaining 150mL water to the beaker. This is approximately 0.1*N* solution of KOH.
- Now, fill up burette with 0.1*N* HCl (secondary standard) and mount it in the burette stand.
- Pipette out 10mL of the approximately 0.1*N* KOH into a 250mL conical flask and add few drops of phenolphthalein indicator. The color of the solution will become pink.
- Titrate against 0.1*N* HCl up to an end point of disappearance of pink color.
- Note down the burette readings and find out the volume of 0.1*N* HCl used in the titration.
- Repeat the titration until getting concordant titer values.
- Now, using the relationship, $V_1N_1 = V_2N_2$, find out the normality of the KOH solution prepared.
- Calculate the volume of water to be added in case of the normality is more than 0.1*N* or the amount of KOH to be added in case the normality is less than 0.1*N*.
- Add either the volume of water or the amount of KOH, as the case may be and mix it thoroughly by stirring.
- Repeat the titration and calculation processes until the normality of the KOH prepared reaches 0.1*N*.

### 8.2.4 Observation and Calculation

| | | |
|---|---|---|
| Volume of HCl consumed in the titration ($V_1$) | = | X mL |
| Volume of 0.1*N* KOH taken ($V_2$) | = | 10mL |
| Normality of HCl ($N_1$) | = | 0.1*N* |
| Normality of the KOH prepared ($N_2$) | = | Y (*N*) |
| We know that, **$V_1N_1 = V_2N_2$ (Law of Volumetric Analysis)** | | |
| $X \times 0.1$ | = | $10 \times Y$ |
| $N_2 = Y$ | = | $(X \times 0.1) / 10$ |

### 8.2.5 Format of Titration Table

0.1 *N* HCl vs Approximately 0.1 *N* KOH

| Sl.No | Aliquot taken in the pipette (mL) | Burette readings (mL) | | Volume of titrant used (mL) | Indicator | End point |
|---|---|---|---|---|---|---|
| | | Initial | Final | | | |
| | | | | | | |
| | | | | | | |
| | | | | | | |

# 9 Working Principle of Some Important Instruments

## *(Smith, 1990; Robinson et al., 2014)*

## 9.1 PH METER

A pH meter is a specific type of voltmeter with a very high impedance of the input channels. The high impedance is a necessary part of the equipment because of high resistance of the pH glass electrode typically used with pH meters (usually between 20 and 100MΩ).

The first commercial pH meters were built around 1936 by Dr. Arnold Beckman in the US and by Radiometer in Denmark. Dr. Beckman's invention helped him launch the successful Beckman Instruments Company.

pH meters work on the Nernst equation and pH is defined as the negative logarithm of hydrogen ion concentration.

$$pH = -\log(H^+)$$

The common pH meter has several inputs for indicator (ion-sensitive or redox) and reference electrode and temperature sensors such as thermocouples.

Two electrodes are used in the determination of pH. One is the reference electrode, which provides standard voltage. It contains two layers, the first layer consists of a strip of platinum sealed in glass and dipped into paste of calomel ($Hg_2Cl_2$, 0.1M) and the second layer is filled with saturated KCl solution. The second electrode is a glass electrode that consists of a high-resistance glass tube with a thin, low-resistance glass bulb at the bottom. It encloses silver-coated silver wire with wax insulation.

When the electrodes are dipped in the solution, the saturated solution of KCl comes out of the reference electrode through the small holes and forms an invisible ionic bridge between electrodes through which current passes. The H ions are absorbed by the glass electrode and electric potential develops between the electrodes. This potential difference is measured in terms of pH by suitable galvanometer.

The pH scale of the device should be calibrated by at least two buffer solutions. Usually one of the buffers used for calibration has pH 7.00 and the second is selected depending on the range where the measurements are to be taken—9.20 for basic

DOI: 10.1201/9781003430100-9

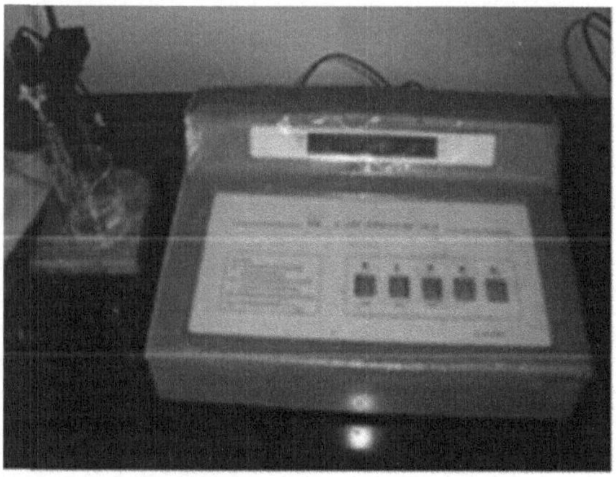

**FIGURE 9.1** pH meter.

solutions and 4.01 for acidic solutions. This correlates the measured potential of the indicator electrode with the pH scale.

General tips:

- The electrode should always remain immersed in the water.
- Allow the instrument to stabilize for a period of at least 30 minutes before use.
- Change the water in the beaker daily.
- If there is strong decline or fluctuation in the reading, fill the electrode with saturated KCl solution.

## 9.2 CONDUCTIVITY METER

A conductivity meter measures the ionic conductivity (or conversely, the resistance) of a liquid. The number it gives cannot be directly related to hardness, but rather, the total ion content of the liquid. What the device usually consists of is a probe which usually has two platinum electrode plates parallel to each other and separated by some small distance. The meter is, in its simplest form, a Wheatstone bridge apparatus with a small oscillator and a readout. The oscillator generates small amplitude (about 5–10mV peak-to-peak) sin wave. The meter is zeroed by dialing in a resistance on one side of the Wheatstone bridge.

The resistance dialed in is then the resistance between the two plates in the probe. Relating this value to a calibration curve for the probe will give you the concentration of ionic species in solution.

### 9.2.1 Reporting the Levels

Conductivity is customarily reported in micro-mhos per centimeter (μmho/cm). In the international system of units, the reciprocal of the ohm is the siemens

(S) and the conductivity is reported as millisiemens or decisiemens per meter (mS/m or dS/m).

$$\text{Conductivity } (\mu S/cm) \times 0.5 = \text{TDS mg/L as NaCl}$$

### 9.2.2 Conductivity Standard Solutions

Conductivity is measured against standard solution of KCl and the following concentrations are used for the calibration:

| Solution type | Mass of KCl/liter | Conductivity at 25°C | Temp. coefficient at 25°C |
|---|---|---|---|
| 1.0M | 71.1352g | 111.3mS/cm | 1.89% |
| 0.1M | 7.4191g | 12.85mS/cm | 1.90% |
| 0.01M | 0.7452g | 1408μS/cm | 1.94% |
| 0.001M | 0.0753g | 146.1μS/cm | 2.04% |

The above table also can be used to calculate the conductivity at any other ambient temperature.

Example: Calculation of conductivity of 0.01*M* KCl at 30°C

$$\begin{aligned} C\,30°C &= C\,25°C + C\,25°C \times 1.94/100 \times (30°C - 25°C) \\ &= 1408\ \mu S + 1408 \times 1.94/100 \times 5 \\ &= 1544.5\ \mu S/cm \end{aligned}$$

## 9.3 UV-VISIBLE SPECTROPHOTOMETER

The instrument used in ultraviolet-visible spectroscopy is called an ultraviolet-visible spectrophotometer. To obtain absorption information, a sample is placed in the spectrophotometer and ultraviolet or visible light at a certain wavelength, or range of wavelengths, is transmitted through the sample. The spectrophotometer measures how much of the light is absorbed by the sample.

The functioning of this instrument is relatively straightforward. A beam of light/radiation from a visible and/or ultraviolet light source (colored red) is separated into its respective component wavelengths by using a prism or diffraction grater. Each monochromatic single wavelength light beam, in turn, is split into two equal intensity beams by using a half-mirrored device. One beam, the sample beam (colored magenta), passes through a small transparent container called a "cuvette" containing a solution of the compound being investigated/studied in a transparent solvent. The other beam, the reference (colored blue), passes through an identical cuvette containing only the solvent.

The intensities of these light beams are then measured by electronic detectors and compared. The intensity of the reference beam, which should have suffered little or no light absorption, is defined as $\mathbf{I_0}$. The intensity of the sample beam is defined as **I**. Over a short period of time, the spectrometer automatically scans all the component

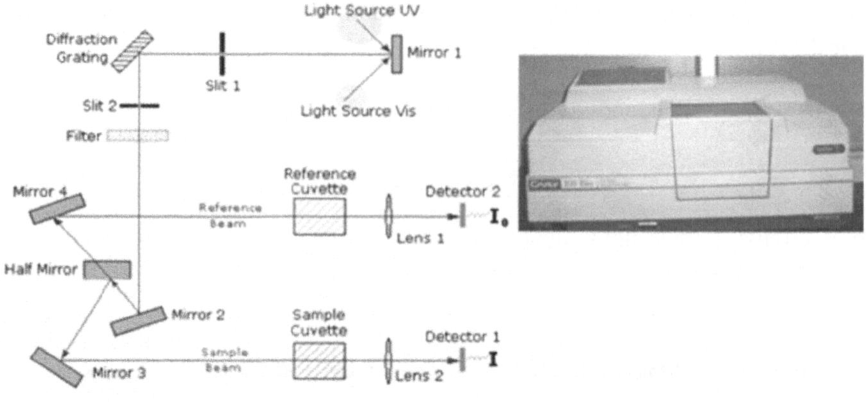


**FIGURE 9.2** UV-visible spectrophotometer and its flow diagram.

wavelengths in the manner described. The ultraviolet (UV) region scanned is normally 200–400nm, and the visible portion is 400–800nm. A diagram of the components of a typical spectrometer is shown in Figure 9.2.

If the sample compound does not absorb the light of a given wavelength, $\mathbf{I} = \mathbf{I}_0$. However, if the sample compound absorbs light, then **I** is less than $\mathbf{I}_0$, and this difference may be plotted on a graph versus wavelength, as shown in Figure 9.2.

Absorption may be presented in terms of transmittance ($T = \mathbf{I}/\mathbf{I}_0$) or absorbance ($A = \log \mathbf{I}_0/\mathbf{I}$). If there is no absorption, T = 1.0 and $A = 0$. Most of the spectrometers display the results of absorbance on the vertical axis, and the commonly observed range is from 0 (100% transmittance) to 2 (1% transmittance). The wavelength of maximum absorbance is a characteristic value, termed as $\lambda_{max}$. The compounds of various nature will have very different absorption maxima and absorbance.

Intensely absorbing compounds must be examined in dilute solution, so that significant light energy is received by the detector, and this requires the use of completely transparent (non-absorbing) solvents. The most commonly used solvents are water, ethyl alcohol, hexane and cyclohexane. The solvents with double or triple bonds, or heavy atoms (e.g., S, Br, and I) are generally avoided. Because the absorbance of a sample will be proportional to its molar concentration in the sample cuvette, a corrected absorption value known as the **molar absorptivity** is used when comparing the spectra of different compounds. This is defined by the Beer-Lambert Law, which states that within small ranges, the concentration of the desired compound varies linearly with the absorbance. Thus UV/vis spectroscopy can determine the concentration of an unknown solution, based on reference molar extinction coefficients or more accurately, using a calibration curve.

$$\textbf{Molar absorptivity, } \varepsilon = \frac{A}{\mathbf{c} \times \mathbf{l}}$$

(where $A$ = absorbance, **c** = sample concentration in moles/liter and **l** = length of light path through the cuvette in cm).

UV/vis spectroscopy is routinely used in the quantitative estimation of solutions of transition-metals and highly conjugated organic compounds. It is possible to do so because transition metals are often colored because of the possibility of d-d electronic transitions within the metal atoms. Organic molecules, especially those with a high degree of conjugation also absorb light in the UV or visible regions of the electromagnetic spectrum.

## 9.4 FLAME PHOTOMETER

Flame photometers use atomic emission for the routine detection of metal salts, principally sodium (Na), potassium (K), lithium (Li), and calcium (Ca). A flame photometer is an instrument used for measuring the spectral intensity of metals present in the metallic salt.

Flame photometry, now more properly called *flame atomic emission spectrometry*, is a relatively old instrumental analysis method. Its origins date back to Bunsen's flame color tests for the *qualitative* identification of select metallic elements. As an analytical method, atomic emission is a fast, simple, and sensitive method for the determination of trace metal ions in solution. Because of the very *narrow* (*c.* 0.01nm) *and characteristic* emission lines from the gas-phase atoms in the flame plasma, the method is relatively free of interference from other elements.

### 9.4.1 THEORY

1. Sample solution is sprayed (aspirated) as fine mist into flame. Conversion of sample solution into an aerosol by atomizer.
2. The heat of the flame vaporizes the sample constituents.
3. By heat of the flame in combination with the action of the reducing gas (fuel), molecules and ions of the sample species are decomposed and reduced to give their respective atoms.
   E.g., $Na^+ + e^- \rightarrow Na$
4. The heat of the flame causes the excitation of some atoms into higher electronic (excited) states.
5. The excited atoms revert to ground state by emission of light energy, of characteristic wavelength; measured by detector.

Atoms in the vapor state gives rise to the line spectra (not band spectra, because no covalent bonds, as there is no vibrational sub-levels to cause broadening).

The colored glass filter is capable of isolating the line of analyte element, if it is well separated from other emission lines. For example, to measure sodium and potassium separately in samples containing both.

```
Na   I I
K              I I
-------------------------------------------
400  500  600  700  800  (λnm)
```

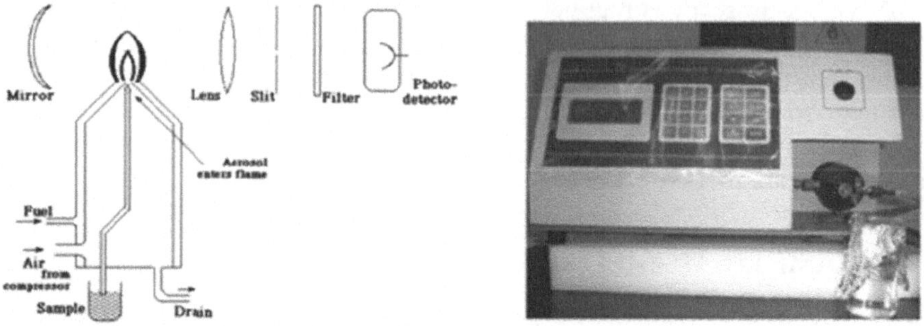


**FIGURE 9.3** A flame photometer and its flow diagram.

### 9.4.2 General Tips for Flame Photometry

- Propane-air or natural gas-air gives good flame—strong heat, minimal background light emission. But always run a solvent blank for setting zero emission.
- Solutions diluted to fall within linear part of emission curve can calibrate with standards accordingly (e.g., from 0.05–0.25mM $Na^+$).
- Use of very low conc. $Na^+$ and $K^+$ solutions → problems of avoiding contamination. Especially $Na^+$ leaches slowly from glass, contact with skin.
- Anion and cation interference effects can cause errors (enhancement or suppression). "Radiation buffer" for dilution of standards and samples to swamp out inconsistencies.
- Internal standard (lithium) useful to counter random flame instability and random dilution errors.

## 9.5 ATOMIC ABSORPTION SPECTROPHOTOMETER

An atomic absorption spectrophotometer (AAS) is a piece of analytical equipment based on atomic absorption spectrophotometry and is used to measure metals in the sample. When a sample is aspirated into the instrument, it is subjected to a heavy thermal environment and, as a result, "the ground state" atoms absorb light energy of a specific wavelength and enter in to the excited state. As the number of atoms in the light path increases, the amount of light absorption also increases. By measuring the amount of light absorbed, a quantitative determination of the amount of analyte element present in the sample can be made.

Every absorption spectrometer have five basic components, which are

- Source of light (cathode lamp)
- A sample cell (absorption cell)
- Monochromator
- Detector
- Output unit

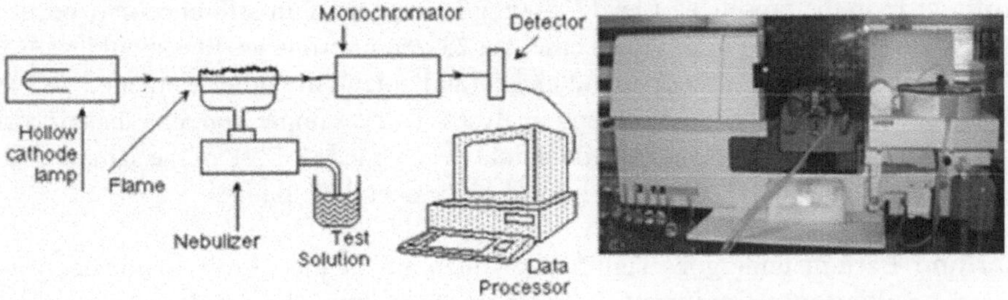


**FIGURE 9.4** Atomic absorption spectrophotometer and its flow diagram.

Absorption in the flame is by vapor phase atoms, giving absorption line spectra. A continuous spectrum light source, even with a high-quality monochromator cannot achieve sufficiently narrow band pass width for absorption line spectra. Therefore, special lamps are used, each emitting line spectrum matched to the line spectrum of the analyte atoms in the flame. The type of lamp is a hollow cathode lamp. A different lamp for each analyte element is required, but some multi-element lamps also available.

### 9.5.1 Make Up Three Standards

The first one should be at the top of the linear range. The concentration of the second standard should be approximately three times the concentration of the first. The concentration of the third standard should be approximately six times the concentration of the first standard.

### 9.5.2 Characteristic Concentration vs Detection Limit

Characteristic concentration in atomic absorption (sometimes called "sensitivity") is defined as the concentration of an element (expressed in ppm or mg/L) required to produce a signal of 1% absorption (0.0044 absorbance units). As long as the measurements are made in the linear working range, the characteristic concentration can be determined by reading the absorbance produced by a known concentration of the element, and therefore,

$$\text{Characteristic Concentration} = \text{Conc. of Std. (ppm or mg/L)} \times 0.\ 0044$$

The characteristic concentration check value is the concentration of element (in mg/L) that will produce a signal of approximately 0.2 absorbance units under optimum conditions at the wavelength listed.

### 9.5.3 Specific Interference Problems in Elemental Analysis by AAS

**Aluminum:** Aluminum may be as much as 15% ionized in a nitrous oxide/acetylene flame. Use of an ionization suppressor (1000μg/mL K as KCl) will eliminate this interference.

**Antimony:** In the presence of lead (1000mg/L), a spectral interference may occur at the 217.6nm resonance line. In this case, the 231.1nm resonance line should be used. Excess concentrations of copper and nickel (and potentially other elements), as well as acids, can interfere with antimony analyses. If the sample contains these matrix types, either matrices of the standards should be matched to those of the sample or the sample should be analyzed using a nitrous oxide/acetylene flame.

**Barium:** Barium undergoes significant ionization in the nitrous oxide/acetylene flame, resulting in a significant decrease in sensitivity. All samples and standards must contain 2mL of the KCl ionization suppressant per 100mL of solution. In addition, high hollow cathode current settings and a narrow spectral band pass must be used because both barium and calcium emit strongly at barium's analytical wavelength.

**Beryllium:** Concentrations of Al greater than 500ppm may suppress beryllium absorbance. The addition of 0.1% fluoride has been found effective in eliminating this interference. High concentrations of magnesium and silicon cause similar problems and require the use of the method of standard additions.

**Calcium:** All elements forming stable oxyanions will complex calcium and interfere unless lanthanum is added. Addition of lanthanum (0.1–1%) to prepared samples rarely presents a problem because virtually all environmental samples contain sufficient calcium to require dilution to be within the linear range of the method.

**Chromium:** Ionization interference may occur if the samples have significantly higher alkali metal content than the standards. If this interference is encountered, an ionization suppressant (KCl) should be added to both samples and standards.

**Magnesium:** All elements forming stable oxyanions (P, B, Si, Cr, S, V, Ti, Al, etc.) will complex magnesium and interfere unless lanthanum is added. Addition of lanthanum to prepared samples rarely presents a problem because virtually all environmental samples contain sufficient magnesium to require dilution.

**Molybdenum:** Interference in an air/acetylene flame from Ca, Sr, $SO_4$, and Fe are severe. These interferences are greatly reduced in the nitrous oxide flame and by the addition of 1000mg/L aluminum to samples and standards.

**Nickel:** High concentrations of iron, cobalt, or chromium may interfere, requiring either matrix matching or use of a nitrous-oxide/acetylene flame. A non-response line of Ni at 232.14nm causes non-linear calibration curves at moderate to high nickel concentrations, requiring sample dilution or use of the 352.4nm line.

**Osmium:** Due to the volatility of osmium, standards must be made on a daily basis, and the applicability of sample preparation techniques must be verified for the sample matrices of interest.

**Potassium:** In air/acetylene or other high temperature flames (>2800°C), potassium can experience partial ionization, which indirectly affects absorption sensitivity. The presence of other alkali salts in the sample can reduce ionization and thereby enhance analytical results. The ionization-suppressive effect of sodium is small if the ratio of Na to K is under 10. Any enhancement due to sodium can be stabilized by adding excess sodium (1000µg/mL) to both sample and standard solutions. If more stringent control of ionization is required, the addition of cesium should be considered.

**Silver:** Since silver nitrate solutions are light sensitive and have the tendency to plate silver out on the container walls, they should be stored in dark-colored bottles. In addition, it is recommended that the stock standard concentrations be kept below 2ppm and the chloride content increased to prevent precipitation. If precipitation is occurring, a 5:2% HCl: $HNO_3$ stock solution may prevent precipitation. Daily standard preparation may also be needed to prevent precipitation of silver.

**Strontium:** Chemical interference caused by silicon, aluminum, and phosphate are controlled by adding lanthanum chloride. Potassium chloride is added to suppress the ionization of strontium. All samples and standards should contain 1mL of lanthanum chloride/potassium chloride solution per 10mL of solution.

**Vanadium:** High concentrations of aluminum or titanium, or the presence of Bi, Cr, Fe, acetic acid, phosphoric acid, surfactants, detergents, or alkali metals, may interfere. The interference can be controlled by adding 1000mg/L aluminum to samples and standards.

**Zinc:** High levels of silicon, copper, or phosphate may interfere. Addition of strontium (1500mg/L) removes the copper and phosphate interference.

### 9.5.4 Instrument Setting for AAS

| Element | Wave length (nm) | Slit width (nm) | Working range (µg/mL) | Sensitivity (µg/mL) | Lamp current (mA) | Flame type |
|---|---|---|---|---|---|---|
| Al | 309.3 | 0.5 | 25–135 | 1.0 | 25.0 | $N_2O$-$C_2H_2$ |
| Al | 396.2 | 0.5 | 25–110 | 1.0 | 25.0 | $N_2O$-$C_2H_2$ |
| As | 193.7 | 0.7 | 1–100 | 45.0 | 300.0 | MHS |
| Ca | 422.7 | 0.5 | 1–4 | 0.02 | 10.0 | $N_2O$-$C_2H_2$ |
| Ca | 422.7 | 0.5 | 1–10 | 0.09 | 10.0 | Air-$C_2H_2$ |
| Cd | 228.8 | 0.5 | 0.5–5 | 0.03 | 15.0 | Air-$C_2H_2$ |
| Co | 240.7 | 0.2 | 1–20 | 0.2 | 6.0 | Air-$C_2H_2$ |
| Cr | 357.9 | 0.2 | 2–20 | 0.2 | 25.0 | Air-$C_2H_2$ |
| Cr | 425.4 | 0.2 | 7–40 | 0.5 | 25.0 | Air-$C_2H_2$ |
| Cu | 324.7 | 0.5 | 1–20 | 0.1 | 15.0 | Air-$C_2H_2$ |
| Cu | 217.9 | 0.2 | 7.5–30 | 0.2 | 15.0 | Air-$C_2H_2$ |
| Fe | 248.3 | 0.2 | 2–20 | 0.1 | 30.0 | Air-$C_2H_2$ |
| Fe | 372.0 | 0.2 | 20–80 | 0.5 | 30.0 | Air-$C_2H_2$ |

| Element | Wave length (nm) | Slit width (nm) | Working range (μg/mL) | Sensitivity (μg/mL) | Lamp current (mA) | Flame type |
|---|---|---|---|---|---|---|
| Hg | 253.7 | 0.7 | 1–200 | 4.2 | 150.0 | MHS |
| K | 766.5 | 0.5 | 1–10 | 0.01 | 6.0 | Air-$C_2H_2$ |
| Mg | 285.2 | 0.5 | 0.1–2 | 0.01 | 6.0 | Air-$C_2H_2$ |
| Mg | 202.6 | 1.0 | 5–20 | 0.1 | 6.0 | Air-$C_2H_2$ |
| Mn | 279.5 | 0.2 | 1–10 | 0.06 | 5.0 | Air-$C_2H_2$ |
| Mn | 403.1 | 0.2 | 7–27 | 0.2 | 5.0 | Air-$C_2H_2$ |
| Mo | 313.3 | 0.2 | 10–50 | 0.8 | 7.0 | $N_2O$-$C_2H_2$ |
| Na | 589.0 | 0.2 | 0.03–1 | 0.02 | 8.0 | Air-$C_2H_2$ |
| Ni | 232.0 | 0.2 | 2–20 | 0.2 | 25.0 | Air-$C_2H_2$ |
| Ni | 352.4 | 0.5 | 6–30 | 0.2 | 25.0 | Air-$C_2H_2$ |
| Pb | 283.3 | 0.5 | 4–40 | 0.2 | 440.0 | Air-$C_2H_2$ |
| Se | 196.1 | 1.0 | 45–180 | 1.0 | 10.0 | $N_2O$-$C_2H_2$ |
| Si | 251.6 | 0.2 | 20–200 | 2.0 | 15.0 | $N_2O$-$C_2H_2$ |
| Zn | 213.9 | 0.5 | 0.5–5 | 0.03 | 15.0 | Air-$C_2H_2$ |

## 9.6 INDUCTIVELY COUPLED PLASMA (ICP) EMISSION SPECTROSCOPY

Inductively coupled plasma (ICP) emission spectroscopy is an atomic emission technique using an argon plasma as an excitation source. However, the design of the source is completely different from the direct-current plasma (DCP). The sample is again introduced into a premix spray chamber, where it is directed up the central tube of the ICP "torch." The torch consists of the concentric tubes with independent argon streams flowing through each. The top of the torch is centered within a radio frequency (RF) induction coil, which is the source of energy for the system. After the ignition, the plasma is propagated through inductive coupling with the RF field generated from the coil. Unlike DCP, there are no electrodes to maintain and replace.

Further the ICP torch is designed specifically to promote penetration of the plasma skin by the sample, allowing sample atoms to experience the full energy of the plasma source. The high temperatures provided by the ICP provide excellent sensitivities for refractory elements and also essentially eliminate chemical interferences. Like all emission techniques, there are no source lamps. By monitoring several wavelengths, either all at once or in a programmed sequence, many elements can be determined in one automated analysis. ICP emission, therefore, offers significant speed advantages over atomic absorption for multielement analyses. Except for the refractory elements, which may be substantially better than even graphite furnace AA, ICP detection limits are comparable to the flame atomic absorption. The high temperatures of the ICP carry one disadvantage. The plasma is so effective in generating excited state species that the rich emission spectra produced increase the probability of spectral interferences.

High-resolution monochromators and sophisticated software for background and inter-element correction are used to deal with this potential problem. Another limitation of ICP emission is the initial cost of the instrumentation. The price for basic ICP systems starts at about the same level as the prices for top-of-the-line automated AA systems. More sophisticated instrumentation can cost two to four times the price of basic systems.

## 9.7 INDUCTIVELY COUPLED PLASMA–MASS SPECTROMETRY (ICP-MS)

Inductively coupled plasma–mass spectrometry (ICP-MS) is one of a growing number of "hyphenated techniques," where the output of one technique becomes the input of another. For ICP-MS, the ICP is used as the ion source for a mass spectrometer. The ions are then spatially separated according to their mass and charge, and measured individually. The major attractiveness of ICP-MS is its exceptional sensitivity combined with high analysis speed. For most elements, ICP-MS offers detection limits that are comparable to or better than those of graphite furnace AA. But ICP-MS can determine many elements in the time required for the determination of one element by graphite furnace AA. ICP-MS also offers the ability for isotopic analysis.

As with the other techniques, ICP-MS also has its limitations. The relative newness of ICP-MS means, while the required instrumentation is well developed, many developments in analytical methodology are yet to be made. This translates into additional effort for the analyst in adapting the technique to their particular analytical needs. Since ICP-MS is not a spectroscopic technique, spectral interferences do not occur. Interferences from mass overlaps due to other isotopes and polyatomic species do occur, however, and may provide erroneous results unless properly corrected.

The major limitation of ICP-MS at this time may be its cost. ICP-MS systems typically are two to four times as expensive as basic ICP emission systems. However, the unique abilities of ICP-MS to provide graphite furnace detection limits with the analytical speed of ICP emission and to perform isotopic analysis capabilities frequently provide the justification needed to overcome cost limitations.

## 9.8 MICROWAVE DIGESTION SYSTEM

A microwave digestion system is an advanced and highly sophisticated system for sample preparation, which utilizes electromagnetic radiation to achieve a higher temperature for the reaction and provides high performance, reliable quality, and unrivaled safety, which is required in sample preparation in order to achieve superior analytical results. Its closed vessel technique helps to speed up reactions by allowing higher temperatures, while preventing the loss of volatile analytes. The resulting low reagent consumption saves time and money and also helps to minimize exposure to corrosive gases and hazardous solvent vapors. The magnetron of the system generates microwaves of 0.3mm–1.0m wavelength having frequencies of 100GHz to 300MHz, which has strong penetrating power on the matrix to extract the analyte of interest.

This instrument is now widely used in domestic, commercial, and scientific segments for a variety of purposes. In the field of research, it is specifically employed

**FIGURE 9.5** Microwave digestion system and its rotor.

in digesting of various abiotic and biotic matrices to extract organic and inorganic chemicals.

## 9.9 AUTO KJELDAHL NITROGEN ANALYZER

Nitrogen determination has a long history in the area of analytical chemistry. Johan Kjeldahl first introduced the Kjeldahl nitrogen method in 1883. While studying proteins during malt production, he developed a method of determining nitrogen content that was faster and more accurate than any method available at the time.

Since 1883, the Kjeldahl method has gained wide acceptance and is now used for a variety of applications. Kjeldahl nitrogen determinations are performed on food and beverages, meat, feed, grain, waste water, soil, and many other samples. The method has been refined and tested for a wide variety of substances and approved by various scientific associations including AOAC, AACC, EPA, ISO, and USDA.

The auto Kjeldahl nitrogen analyzer consists of two separate units, block digester and distillation, titration assembly. In block digester digestion is performed at 420$^0$C further digested samples transferred into distillation titration assembly where liberated ammonia collected in boric acid solution and titrated against standard acid. After the end of process result can be read out on display.

The Kjeldahl method may be broken down into three main steps:

**Digestion**—the decomposition of nitrogen in organic samples utilizing a concentrated sulfuric acid. The end result is an ammonium sulfate solution.

$$\text{Organic N} + H_2SO_4 \rightarrow (NH_4)_2SO_4 + H_2O + CO_2 + \text{other sample matrix by-products.}$$

**Distillation**—adding excess base to the acid digestion mixture to convert $(NH_4)_2SO_4$ to $NH_3$, followed by boiling and condensation of the $NH_3$ gas in a receiving solution.

**FIGURE 9.6** Automatic nitrogen analyzer with block digester.

$$(NH_4)_2SO_4 + 2NaOH \rightarrow 2NH_3 + Na_2SO_4 + 2H_2O$$

$$NH_3 + H_3BO_3 \rightarrow NH_4^+ : H_2BO_3^- + H_3BO_3$$

**Titration**—to quan solution, the amount of nitrogen in a sample can be calculated from the quantified amount of ammonia ions in the receiving solution.

$$NH_4^+ : H_2BO_3^- + HCl \rightarrow NH_4Cl + H_3BO_3$$

$$N(\%) = \frac{(S\text{-}B) \times N \times 1.407}{\text{Sample weight (g)}}$$

Where
S = volume of acid used against sample.
B = volume of acid used against blank.
N = normality of acid.

If it is desired to determine protein percentage instead of nitrogen percentage, the calculated nitrogen percentage is multiplied by a factor, the magnitude of the factor depending on the sample matrix.

Many protein factors have been developed by AACC and AOAC for use with various types of samples such as, 6.38 for milk and dairy, 5.95 for rice, 5.70 for wheat flour, and 6.25 for other grains.

## 9.10 HIGH-PERFORMANCE LIQUID CHROMATOGRAPHY (HPLC)

HPLC is a form of liquid chromatography to separate compounds that are dissolved in solution. HPLC is a popular method of analysis because it is easy to learn and use. It is not limited by the volatility or stability of the sample compound.

High pressure liquid chromatography was developed during the 1970s and quickly improved with the development of column packing materials and the additional convenience of online detectors. In the late 1970s, new methods including reverse phase liquid chromatography allowed for improved separation between very similar compounds. Modern HPLC has many applications including separation, identification, purification, and quantification of various compounds.

Although HPLC is widely considered to be a technique mainly for biotechnological, biomedical, and biochemical research as well as for the pharmaceutical industry, these fields currently comprise only about 50% of HPLC users. Currently HPLC is used by a variety of fields including cosmetics, energy, food, and environmental industries.

HPLC instruments consist of a reservoir of mobile phase, a pump, an injector, a separation column, and a detector. Compounds are separated by injecting a plug of the sample mixture onto the column. The different components in the mixture pass through the column at different rates due to differences in their partitioning behavior between the mobile liquid phase and the stationary phase.

The mobile phase in HPLC refers to the solvent being continuously applied to the column, or stationary phase. The mobile phase acts as a carrier for the sample solution. A sample solution is injected into the mobile phase of an assay through the injector port. As a sample solution flows through a column with the mobile phase, the components of that solution migrate according to the non-covalent interactions of the compound with the column. There are several types of mobile phases, these include isocratic, gradient, and polytypic.

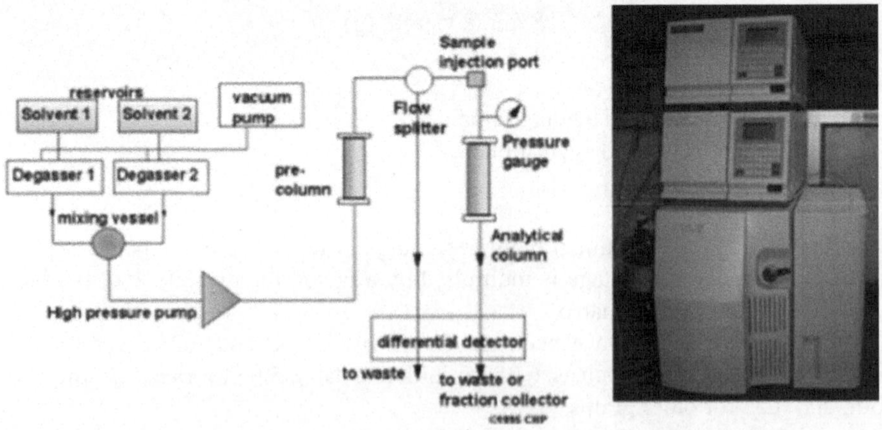


**FIGURE 9.7** High performance liquid chromatograph and its flow diagram.

In isocratic elution compounds are eluted using constant mobile phase composition. In gradient elution different compounds are eluted by increasing the strength of the organic solvent. The sample is injected while a weaker mobile phase is being applied to the system. The strength of the mobile phase is later increased in increments by raising the organic solvent fraction, which subsequently results in elution of retained components. This is usually done in a stepwise or linear fashion. Polytypic mobile phase, sometimes referred to as mixed-mode chromatography, is a versatile method in which several types of chromatographic techniques, or modes, can be employed using the same column. These columns contain rigid macro-porous hydrophobic resins covalently bonded to a hydrophilic organic layer. SEC, IEC, hydrophobic, or affinity chromatography are some of the methods that may be utilized. By changing the mobile phase, the mode of separation is thereby changed which allows the chromatographer to achieve the desired selectivity in the separations.

The stationary phase in HPLC refers to the solid support contained within the column over which the mobile phase continuously flows. The sample solution is injected into the mobile phase of the assay through the injector port. As the sample solution flows with the mobile phase through the stationary phase, the components of that solution will migrate according to the non-covalent interactions of the compounds with the stationary phase.

The chemical interactions of the stationary phase and the sample with the mobile phase determine the degree of migration and separation of the components contained in the sample. Columns containing various types of stationary phases are commercially available. Some of the more common stationary phases include liquid-liquid, liquid-solid (adsorption), size exclusion, normal phase, reverse phase, ion exchange, and affinity.

Samples are injected into the HPLC via an injection port. The injection port of an HPLC commonly consists of an injection valve and the sample loop. The sample is typically dissolved in the mobile phase before injection into the sample loop. The sample is then drawn into a syringe and injected into the loop via the injection valve. A rotation of the valve rotor closes the valve and opens the loop in order to inject the sample into the stream of the mobile phase. Loop volumes can range between 10μl and over 500μl. In modern HPLC systems, the sample injection is typically automated.

There are various types of pumps available for use with HPLC analysis, they are:

- Reciprocating piston pumps
- Syringe type pumps
- Constant pressure pumps

The detector for an HPLC is the component that emits a response due to the eluting sample compound and subsequently signals a peak on the chromatogram. It is positioned immediately posterior to the stationary phase in order to detect the compounds as they elute from the column. The bandwidth and height of the peaks may usually be adjusted using the coarse and fine-tuning controls, and the detection and sensitivity parameters may also be controlled (in most cases). There are many types of detectors that can be used with HPLC. Some of the more common detectors include refractive index (RI), ultraviolet (UV), fluorescent, radiochemical, electrochemical, near-infrared (near-IR), mass spectroscopy (MS), nuclear magnetic resonance (NMR), and light scattering (LS).

Refractive index (RI) detectors measure the ability of sample molecules to bend or refract light. This property for each molecule or compound is called its refractive index. For most RI detectors, light proceeds through a bi-modular flow-cell to a photodetector. One channel of the flow-cell directs the mobile phase passing through the column while the other directs only the mobile phase. Detection occurs when the light is bent due to samples eluting from the column, and this is read as a disparity between the two channels.

**Ultraviolet (UV) detectors** measure the ability of a sample to absorb light. This can be accomplished at one or several wavelengths.

**Fluorescent detectors** measure the ability of a compound to absorb then re-emit light at given wavelengths. Each compound has a characteristic fluorescence. The excitation source passes through the flow-cell to a photodetector while a monochromator measures the emission wavelengths.

**Mass spectroscopy (MS) detectors**—the sample compound or molecule is ionized, it is passed through a mass analyzer, and the ion current is detected. There are various methods for ionization.

**Nuclear magnetic resonance (NMR) detectors**—certain nuclei with odd-numbered masses, including H and $^{13}C$, spin about an axis in a random fashion. However, when placed between poles of a strong magnet, the spins are aligned either parallel or anti-parallel to the magnetic field, with the parallel orientation favored since it is slightly lower in energy. The nuclei are then irradiated with electromagnetic radiation which is absorbed and places the parallel nuclei into a higher energy state; consequently, they are now in "resonance" with the radiation. Each H or C will produce different spectra depending on their location and adjacent molecules, or elements in the compound, because all nuclei in molecules are surrounded by electron clouds which change the encompassing magnetic field and thereby alter the absorption frequency.

**Light-scattering (LS) detectors**—when a source emits a parallel beam of light that strikes particles in a solution, some light is reflected, absorbed, transmitted, or scattered.

**Near-infrared detectors** operate by scanning compounds in a spectrum from 700nm to 1100nm. Stretching and bending vibrations of particular chemical bonds in each molecule are detected at certain wavelengths.

### 9.10.1 Liquid Chromatography Applications

Liquid chromatography (LC) has been used in an extremely wide range of analytical methods and it is impossible to give a comprehensive set of examples that would illustrate its wide applicability. The following are a few LC analyses that may indicate the scope of the technique and give the reader some idea of its importance and versatility.

An example of the use of reversed phase chromatography (employing a C18 column) for the separation of some pesticides is shown below. The column used was 25cm long and 3mm in diameter, packed with silica-based, C18 reverse phase packing particles, size 5μm. The mobile phase consisted of (A) 2mM sodium acetate (pH 6.5) with 5% acetonitrile and (B) 100% acetonitrile and the flow-rate was 0.35mL/min.

The gradient was 2 min. 10% B; 10 to 70 min. 45% B. Column temperature was 40°C and peaks detected by UV detector on 245nm. The retention time of the last peak is about 80 minutes. This procedure trades efficiency for time and allows the separation to be achieved in the minimum time given the column and phase system that has been chosen.

## 9.11 GAS CHROMATOGRAPHY (GC)

In gas chromatography (GC), the stationary phase is a high-boiling liquid and the mobile phase is an inert gas. The process of gas chromatography is carried out in a specially designed instrument. A very small amount of liquid mixture is injected into the instrument and is volatilized in a hot injection chamber. Then, it is swept by a stream of inert carrier gas through a heated column which contains the stationary, high-boiling liquid. As the mixture travels through this column, its components go back and forth at different rates between the gas phase and dissolution in the high-boiling liquid, and thus separate into pure components. Just before each compound exits the instrument, it passes through a detector. When the detector "sees" a compound, it sends an electronic message to the recorder, which responds by printing a peak on a piece of paper.

The GC consists of an injection block, a column, and a detector. An inert gas flows through the system. The injection chamber is a heated cavity that serves to volatilize the compounds. The sample is injected by syringe into this chamber through a port that is covered by a rubber septum. Once inside, the sample becomes vaporized and is carried out of the chamber and onto the column by the carrier gas.

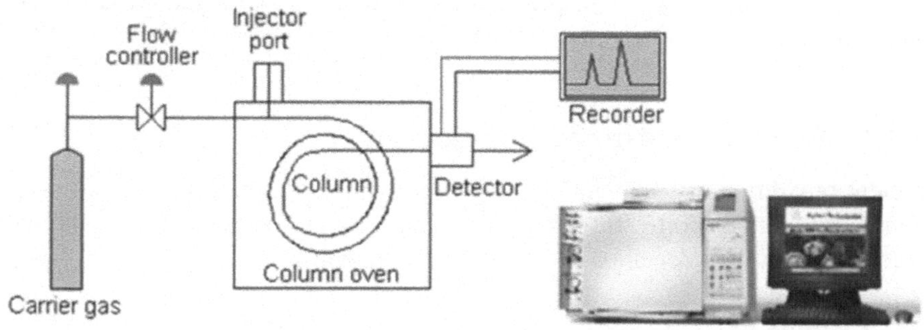


**FIGURE 9.8** Gas chromatography and its flow diagram.

The column is an integral part of the GC system. Inside the column is the important component: the stationary phase composed of the high-boiling liquid. There are two general types of column, *packed* and *capillary* (also known as *open tubular*). Packed columns contain a finely divided, inert, solid support material (commonly based on *diatomaceous earth*) coated with liquid stationary phase. Most packed columns are 1.5–10m in length and have an internal diameter of 2–4mm. Capillary columns have an internal diameter of a few tenths of a millimeter. They can be one of two types; *wall-coated open tubular* (WCOT) or *support-coated open tubular* (SCOT). Wall-coated columns consist of a capillary tube whose walls are coated with liquid stationary phase. In support-coated columns, the inner wall of the capillary is lined with a thin layer of support material such as diatomaceous earth, onto which the stationary phase has been adsorbed. SCOT columns are generally less efficient than WCOT columns. Both types of capillary column are more efficient than packed columns.

The carrier gas is an inert gas, helium. The flow rate of the gas influences how fast a compound will travel through the column; the faster the flow rate, the lower the retention time. Generally, the flow rate is held constant throughout a run.

Each instrument will have a different setting for column temperature, injection port temperature, detector temperature.

There are many detectors that can be used in gas chromatography. Different detectors will give different types of selectivity. A *non-selective* detector responds to all compounds except the carrier gas, a *selective detector* responds to a range of compounds with a common physical or chemical property, and a *specific detector* responds to a single chemical compound. Detectors can also be grouped into *concentration dependent detectors* and *mass flow dependent detectors*. The signal from a concentration dependent detector is related to the concentration of solute in the detector, and does not usually destroy the sample. Dilution with make-up gas will lower the detector's response. Mass flow dependent detectors usually destroy the sample, and the signal is related to the rate at which solute molecules enter the detector.

The response of a mass flow dependent detector is unaffected by make-up gas.

Two devices are used to record the GC traces/areas under peaks, integrating recorders and computer program. Each type of device records the messages sent to them by the detector as peaks, calculates the retention time, and calculates the area under each peak; all of this information is included in the printout. For similar compounds, the area under a GC peak is roughly proportional to the amount of compound injected.

### 9.11.1 Factors That Affect GC Separations

Efficient separation of compounds in GC is dependent on the compounds traveling through the column at different rates. The rate at which a compound travels through a particular GC system depends on the factors listed below:

- **Volatility of compound:** Low boiling (volatile) components will travel faster through the column than will high boiling components.

- **Polarity of compounds:** Polar compounds will move more slowly, especially if the column is polar.
- **Column temperature:** Raising the column temperature speeds up all the compounds in a mixture.
- **Column packing polarity:** Usually, all compounds will move slower on polar columns, but polar compounds will show a larger effect.
- **Flow rate of the gas through the column:** Speeding up the carrier gas flow increases the speed with which all compounds move through the column.
- **Length of the column:** The longer the column, the longer it will take all compounds to elute. Longer columns are employed to obtain better separation.

Generally, the number one factor to consider in separation of compounds on the GCs in the teaching labs is the boiling points of the different components. Differences in polarity of the compounds are only important if you are separating a mixture of compounds that have widely different polarities. Column temperature, the polarity of the column, flow rate, and length of a column are constant in GC runs in organic chemistry teaching labs. For each planned GC experiment, these factors have been optimized to separate your compounds and the instrument set-up by the staff.

## 9.12 DISSOLVED OXYGEN METER

There are two types of DO probes available, including a laboratory probe and a field probe. Both types operate on the same principle. An advantage of the field probe is that it eliminates the need to collect, transport, and then perform analysis on a sample.

Oxygen-sensitive membrane probes contain two solid metal electrodes immersed in an electrolyte solution that allows the passage of an electrical current. The electrode pair is covered with a semi-permeable membrane which allows the passage of DO molecules. When the oxygen molecules pass through the membrane, they cause the electrical current already present to change. This change is reflected by the movement of the needle on the meter.

Allow the instrument to warm up for at least 15 minutes before taking the first reading of the day. Leave the instrument on between measurements. This will eliminate the need for warm-up periods between tests. Calibrate at least once per day. Distilled or deionized water, and not effluent, should be used for the probe calibration. Always stir the sample while taking a reading.

## 9.13 DIRECT CURRENT PLASMA (DCP) EMISSION

DCP is an atomic emission technique. The sample is aspirated into a premix spray chamber through a nebulizer using a system very similar to that for atomic absorption. However, instead of combustible gases, argon is used as a transport gas for the sample. The sample aerosol in a stream of argon is directed at a set of electrodes, across which a high voltage electrical potential is applied. The resulting electrical discharge between the electrodes supplies enough energy to ionize the argon into a "plasma" of positively charged argon ions and free electrons. The thermal energy of the plasma, in turn, atomizes sample constituents and creates excited state atoms, which emit their characteristic atomic emission spectra. DCP was the first plasma technique applied to routine atomic emission analyses. In its early days, it was an especially valuable complement to atomic absorption, in that DCP provided good detection limits for the refractory elements, for which atomic absorption was not particularly sensitive. DCP is also capable of simultaneous multielement analysis and qualitative, as well as quantitative, analysis. DCP carries with it some significant disadvantages, however. The electrodes that form the DC arc are continually eroded and burned away during operation.

This imposes a maintenance problem of continual adjustment and replacement of the electrodes. In addition, the very high temperatures for which plasmas are known are not fully realized in the DCP design. Due to a highly resistant "skin effect," the sample does not penetrate into the hottest part of the plasma, but is instead deflected around it. The analytical measurement normally is made just underneath the hottest part of the plasma, where temperatures are hot enough to provide good sensitivity for refractory elements but not hot enough to eliminate chemical and ionization effects. Procedures for reagent addition are usually prescribed to deal with these interferences. Because these limitations are not normally encountered with inductively coupled plasma (ICP) systems, ICP emission is normally a preferred emission technique.

## 9.14 FIBER ANALYZER

This instrument used for analysis of fiber, such as crude fiber, neutral detergent fiber (NDF), acid detergent fiber (ADF), and acid detergent lignin (ADL).

Fiber analyzer is specifically designed for fiber determination in accordance with the Weende, Van Soest, and other recognized methods. There is single or sequential

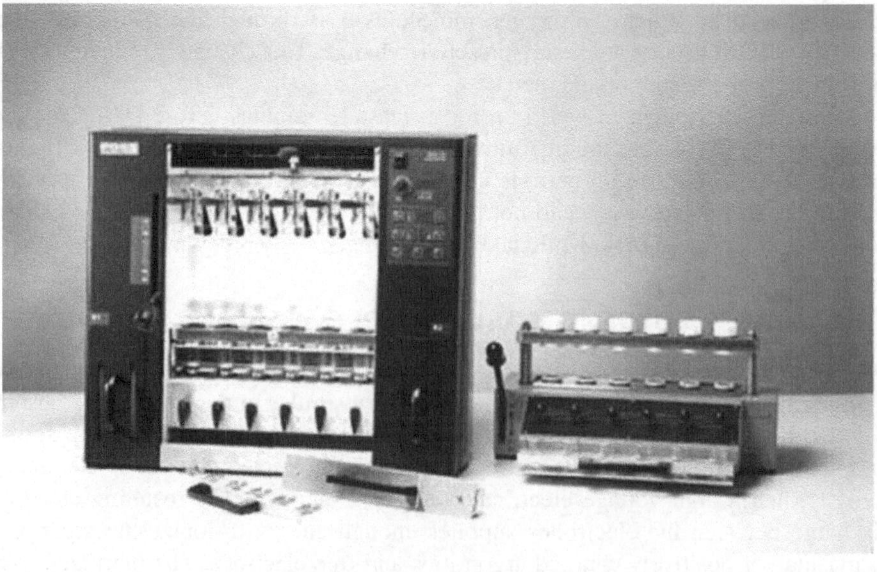

**FIGURE 9.9** Automatic fiber analyzer.

extractions including boiling, use of internally preheated reagents, rinsing and filtration are performed under reproducible and controlled conditions. This system is made of two units, hot extractor and cold extractor. The hot extractor, for hot hydrolysis and extraction, features built-in systems for heating, filtration, automatic preheating, and addition of reagents, while the cold extraction unit is for de-fatting samples, extraction at ambient temperatures (e.g., in lignin determination), and solvent dehydration of fiber residues. Samples are handled in specially designed filter crucibles. Crucibles are used both as an integral part of the assembly during extraction, rinsing and filtration and as sample vessels during weighing, drying, and ashing.

# 10 Collection and Preparation of Soil Samples for Laboratory Analysis

## 10.1 INTRODUCTION

Soil is analyzed to know the nature of soil, to classify and advocate to farmers about the peculiarities of the soils and how much of fertilizers should be applied for better crop production.

A composite soil, of course, cannot be moved to into a laboratory. The value of the laboratory work depends upon care in sampling. Each soil sample needs to be a fair representative of the specific area or horizon worth sampling.

If the sample is to be a representative of an area, it is necessary to take large number of samples spread over the area, pool them, and subsample it so as to get a sample of desired size. For soil survey work, samples are collected from a profile that is typical of the soil of the surrounding area.

## 10.2 MATERIALS REQUIRED

- Spade
- Khurpi
- Auger (screw or posthole type)
- Core sampler
- Soil testing tube (for wet soil)
- Sampling bags and
- Plastic basin or bucket

## 10.3 COLLECTION OF SOIL SAMPLES

### 10.3.1 Collection of Soil Samples from the Field

- Normally each field may be treated as a sampling unit. But two or more fields, which are similar in appearance, production, and past-management practices, may be grouped together into a single sampling unit.
- Samples should be collected separately from areas that differ in soil color or past-management practices such as liming, fertilization, cropping pattern, etc.

DOI: 10.1201/9781003430100-10

- During the collection of soil, avoid dead furrows, old manure or lime piles, wet spots, areas near trees, manure pits, compost pits, and irrigation channels.
- The sampling should be done in a zig-zag pattern across the field to get homogeneity.
- A wise soil sample collecting agent is one who collects samples in the presence of the owner or cultivator of the land who is the best judge in deciding which area of the farm should be sampled separately.
- Scrap away the surface litter and insert the sampling auger to plough depth (15cm).
- Take at least 15 samples randomly distributed over each area and place them in a clean bucket.
- If a sampling auger is not available, make a "V" shaped cut to a depth of 15cm using a spade and remove a 1.0–1.5cm thick slice of soil from top to bottom of the exposed face of the "V" shaped cut and place in a clean bucket.
- Thoroughly mix the samples taken from 15 or more spots of each area.
- Remove foreign bodies such as plant roots, stubbles, pebbles, stones, or gravels.
- By quartering, discard all but 0.5–1kg of soil. Quartering is done by dividing the thoroughly mixed soil into four equal parts and discarding two opposite quarters. Remix the remaining two quarters and again divide into four equal parts and reject the opposite two. Repeat this procedure until to get 0.5–1kg of soil.
- Instead of quartering, the compartmentalization method can also be followed. For this, spread the soil on a clean hard surface and mark lines from both sides and create number of compartments. Take a little quantity of soil from each compartment and put into a clean container. Repeat the process of collection until the required quantity of soil is collected.
- Store the soil in a clean bag or container with proper labeling for further analysis.
- Sampling should be done after the harvest of the crop.
- In case the sampling is necessary during crop growth, sample between lines of growing plants.
- Avoid storing samples in fertilizer bags.

### 10.3.2 Collection of Soil Samples from a Profile

- After the profile has been exposed, clean one face of the pit carefully with a spade and note the succession and depth of each horizon.
- Prick the surface with a knife or edge of the spade to show up the structure, color, and compactness of soil.
- Describe the profile as per the standard terminologies.
- Use a Munsell color chart for noting the color and find out the texture by the feel method.
- Collect samples from each horizon by holding a large basin at the bottom limit of the horizon while the soil above is loosened with a khurpi.
- The sample is mixed and transferred to a bag after labeling.

## 10.4 PREPARATION OF SOIL SAMPLES FOR ANALYSIS

- The soil samples received at the laboratory should be air-dried in shade and spread on a sheet of paper after breaking large lumps, if present, with a wooden mallet.
- The sample is further ground by pounding with a wooden mallet in such a way that the aggregate particles are broken down to ultimate soil particles.
- The soil thus prepared is sieved through a sieve with round holes of 2mm in diameter.
- The material on the sieve is again ground and sieved until all aggregate particles are fine enough to pass through and only stones and organic residues remain on the sieve.
- Mix well the "fine soil" got by sieving and store in a suitable bottle or container with one label on the outside and another inside the container.
- For determination of organic carbon, powder and sieve the soil through a 0.2mm sieve.
- From the soil samples meant for micronutrient analysis, iron, brass, copper, and zinc containers must be avoided for collection and storage of soil samples.

### 10.4.1 Subsampling for Analysis

- The soil in the bottle is emptied on a clean thick sheet of paper and evenly spread with a sampling knife.
- It is heaped into a cone by raising the four ends of the paper.
- It is again mixed well and evenly spread on the paper as before.
- The process is repeated three to four times to ensure uniformity and then finally spread evenly on the paper again.
- Now it is divided into four quarters and small quantity of soil is taken for various points in each quarter to get a representative sample for analysis.

## 10.5 THINGS TO LEARN

- Draw the sampling tools used for the collection of soil samples.
- Tabulate the sampling tools to be used under various soil conditions.
- Illustrate the relationship between the number of samples to be taken and acceptable error, with examples.
- Define the various terminologies associated with collection and preparation of soil samples for laboratory analysis.
- What is the procedure for collecting the soil samples (depth) for horticultural, plantation, and orchard crops?
- What is the sampling depth for problematic soils?

# 11 Estimation of Soil pH

## 11.1 INTRODUCTION

The acidity, neutrality, or alkalinity of a soil is measured in terms of hydrogen ion activity (active concentration) of the soil-water system. The pH is defined as a negative logarithm of hydrogen ion concentration or the log of the reciprocal of the hydrogen ion concentration.

Thus, the pH of a soil is a measure of only the intensity of acidity and not the amount of acid present. The pH scale varies from 0 to 14. Pure water, which is exactly neutral at 25°C, has a pH value of 7. As the pH decreases from 7 to 0, the acidity of the solution increases and as pH increases from 7 to 14, the solution becomes more and more alkaline.

Most agricultural soils have a pH varying from 4.5 to 8.5, although soils with pH as low as 2 or as high as 10 or 11 are also known. The active concentration of hydrogen is termed as "active acidity," which is 1/50000 or 1/100000 times lesser than the "exchangeable acidity" or "exchangeable hydrogen." The soil acidity classes are as follows:

| | | | |
|---|---|---|---|
| Extremely acidic | pH < 4.5 | Neutral | pH 6.6–7.3 |
| Very strongly acidic | pH 4.5–5.0 | Mildly alkaline | pH 7.4–8.0 |
| Strongly acidic | pH 5.1–5.5 | Strongly alkaline | pH 8.1–9.0 |
| Medium acidic | pH 5.6–6.0 | Very strongly alkaline | pH > 9.0 |
| Slightly acidic | pH 6.1–6.5 | | |

## 11.2 IMPORTANCE

- The pH determination is an indispensable means for characterizing soil from the standpoints of nutrient availability, and physical condition *viz.* structure, permeability, etc.
- It provides information on the potency of toxic substances present in the soil.
- It is indicative of the status on microbial environment/community and its net effect on the mineralization of organic residues and/or immobilization of available nutrients.
- Ascertaining the soil pH provides the most rational basis for managing soils for selective crops, pasture cultivation, forestry, etc.

DOI: 10.1201/9781003430100-11

- pH measurement of soil in water and KCl systems provides information on the nature of charge distribution on soil colloids, which will have a far-reaching effect on nutrient management and utilization.
- The pH of soil provides a basis for the management of problematic soils like acid soils, saline soils, alkali soils, etc.

## 11.3 PRINCIPLE

A glass/indicator electrode in contact with hydrogen ions of the solution acquires an electric potential that depends on the concentration of hydrogen ions. This is measured potentiometrically against some reference electrode, which is usually a calomel electrode. The potential difference between glass electrode and calomel electrode is expressed in pH units.

Two electrodes are used in the determination of pH. One is the reference electrode, which provides a standard voltage and is usually a saturated calomel electrode, which has two layers of saturated solution of KCl and mixture of solid $HgCl_2$ and Hg. The outer tube is usually 5–15cm long, 0.5–1cm in diameter. The mixture of solid $HgCl_2$ + Hg paste is contained in an inner tube that is connected to the saturated KCl solution in the outer tube by means of small opening. The resistance developed by this type of electrode is 2000–3000 Ohms.

The outer electrode is glass electrode that consists of a tube enclosing a lead wire made up of Ag coated with AgCl. This wire is again enclosed in a wax insulation. To the tube at the bottom is attached a glass bulb made up of a special kind of glass which is sensitive to hydrogen ions. The thickness of the glass membrane varies from 0.03mm to 0.1mm and has a resistance of 50–500 mega Ohms.

When these electrodes are dipped in solution, the saturated solution of KCl comes out of reference through small holes and forms an invisible ionic bridge between electrodes through which current passes.

The hydrogen ions are absorbed by the glass electrode and, depending upon the amount of hydrogen ions present in the solution, an electric potential develops between electrodes. This potential difference is measured in terms of pH units by suitable galvanometer.

## 11.4 APPARATUS AND MATERIALS REQUIRED

- 100mL beaker
- pH buffer solutions (pH 4.0, 7.0, and 9.2)
- Glass rod
- pH meter

## 11.5 PROCEDURE

### 11.5.1 Standardization of pH Meter

- Switch on the pH meter and allow it warm for ten minutes.
- Keep the pH selector switch in the zero position.

- Set the temperature compensation control to the solution temperature.
- Adjust the zero-adjustment knob so that the pointer in the meter reads exactly zero when the electrodes are immersed in distilled water.
- Lift electrodes from distilled water and wipe dry using filter paper/tissue paper, then dip them in standard buffer solution of known pH (4.0, 7.0, and 9.2).
- Change the function switch to particular pH range (0–7 or 7–14) and adjust the standardization knob till the pointer reads the correct pH value of the buffer solution. Do not disturb the zero-knob adjustment.

### 11.5.2 pH Measurement

- Weigh 20g air-dry soil passed through a 2mm sieve and transfer to a clean 100mL beaker.
- Add 50mL distilled water (1:2.5 ratio of soil water suspension).
- Using a glass rod, stir the content intermittently and allow it to stand for 30 minutes.
- Wash the electrodes carefully with a jet of distilled water and wipe it dry with a piece of filter/tissue paper.
- Stir the soil suspension again just before taking the reading in the pH meter.
- Immerse the electrodes into the beaker containing soil water suspension and change the function to the particular pH range.
- Record the meter reading as pH units.

## 11.6 THINGS TO LEARN

- Differentiate active and exchangeable acidity.
- What is the importance of determining the soil pH?
- Give the pH range of acidic, saline, alkali and saline-alkali soils.

# 12 Estimation of Electrical Conductivity of Soil

## 12.1 INTRODUCTION

The electrical conductivity (EC) of the soil is a measure of soluble salts present in the soil and is expressed as millimhos/cm or decisiemens/m ($dSm^{-1}$).

## 12.2 PRINCIPLE

As the amount of the soluble salts in a solution increases, the electrical conductivity also increases. This electrical conductivity is measured in terms of resistance offered to the flow of current using a conductivity bridge.

It is known that solutions offer some resistance to the passage of current through them, depending upon the concentration of the soluble salts present. Hence, EC is measured in terms of electrical resistance between parallel electrodes immersed in the soil water suspension.

In such a system, the solution between the electrodes becomes the electrical conductor to which the physical laws relating to resistance are applicable. The electrical resistance "R" is directly proportional to the distance "L" between the electrodes and inversely proportional to the cross-sectional area "A" of the conductor.

$$\text{Hence, } R \,\alpha\, L/A \text{ or } R = r \times L/A$$

Where r = proportionality constant known as electrical conductivity

$$\text{If } L = 1\text{cm and } A = 1\text{cm}^2\text{, then } R = r$$

Where r is called specific resistivity

Hence, specific resistance is the resistance of a conductor of 1cm in length and $1cm^2$ in cross sectional area.

The higher the salt content, the higher the passage of current and the lesser the resistance to the flow of the current. Hence, the reciprocal of specific resistivity is called specific conductivity.

DOI: 10.1201/9781003430100-12

Therefore, specific conductivity is defined as the conductivity of a solution enclosed in a cell whose electrodes are of exactly 1cm distance and possess a surface area of $1cm^2$.

The resistance is expressed as Ohms/cm and the electrical conductivity is expressed as mhos/cm as the electrical conductivity is the reciprocal of electrical resistance.

It is not possible to make a conductivity bridge having electrodes $1cm^2$ in area and placed exactly 1cm apart. Hence, the factor called the cell constant is determined for the given cell as a correction factor.

Modern conductivity meters are calibrated to read directly the electrical conductivity/conductance of a given cell.

## 12.3 APPARATUS AND MATERIALS REQUIRED

- 100mL beaker
- Glass rod
- 0.1*N* KCl solution
- Saturated $CaSO_4$ solution
- Conductivity bridge/EC meter

## 12.4 PROCEDURE

- Switch on the EC meter/conductivity bridge and allow it to stand for ten minutes.
- Check the instrument with saturated calcium sulfate solution and 0.1*N* KCl solutions having 2.2 $dSm^{-1}$ and 1.41 $dSm^{-1}$ respectively.
- Use the same soil water suspension used for the pH measurement for the determination of EC.
- Stir the contents and allow the soil to settle for 15 minutes.
- Wash the electrodes carefully and immerse them into soil suspension.
- Adjust the temperature correction and adjust the meter knob until the magic eye of the null indicator is at the widest in width.
- The readings on the scale at this position indicate the electrical conductivity.
- Multiply this value of EC by cell constant usually noted on the cell itself to get specific conductivity.

## 12.5 THINGS TO LEARN

- Differentiate electrical conductivity and specific conductivity.
- What is specific resistivity?
- How the soils can be interpreted based on their electrical conductivity?
- What is the relationship between soluble salts and electrical conductivity?

# 13 Estimation of Organic Carbon Content of Soil *[Titrimetric/Walkley and Black (1934) Method]*

## 13.1 INTRODUCTION

The role of soil organic matter, in relation to soil fertility and physical conditions, is widely recognized. The organic matter is the source of plant nutrients, which are released in assimilable forms during microbial degradation. A major proportion of N (95–99% of the total), P (33–67% of the total) and S (75% of the total) in soils occur in organic combinations, which mineralize to release the nutrients in inorganic forms to be used by plants. Nevertheless, it serves as a reservoir of plant nutrients, in promoting water storage, and in regulating microbial activity.

The organic matter content of a soil varies from 0.344% in very sandy arid soils to more than 86% in peats and mucks. Soil organic matter contains 5% N, and 0.5% each of P and S, thereby, having a N:P:S ratio of 10:1:1. The organic matter content of a soil is estimated by the amount of organic carbon (C) present, as this element represents, on an average, 52–58% of the organic matter, and the C:N ratio is 10–15. In the subsoil, the average organic carbon amount is 36–44% (40%), and the C:N ratio is < 8. In order to find out the amount of humified organic material, the organic carbon content is multiplied by 1.724, which represents the ratio between the humus and the organic carbon (100/58) and is known as the "Van Bemmelen factor."

The organic matter content of soils is estimated from the organic carbon, determined by using the Titrimetric/Walkley and Black Method. The Titrimetric determination, also known as the "wet-digestion method" involves a rapid titration procedure for the estimation of organic carbon content of soils.

## 13.2 IMPORTANCE

- For ordinary routine work, rapid titration method is useful to determine indirectly the N status of soil.
- However, the organic carbon status cannot be taken as a reliable of available N, because the C:N ratio of different soils varies widely.
- The phosphorus and sulfur status of a soil can also be ascertained to a certain extent from such determination.
- This method oxidizes a lower percentage of the total organic C in soils. However, it has been found that, if this method is modified by applying external

DOI: 10.1201/9781003430100-13

heat, immediately adding sulfuric acid to the sample, the oxidation of organic C is nearly complete and the degree of oxidation is more consistent than in the regular method.

- In the regular method, only 77% organic carbon gets oxidized. So, a correction factor should be used in calculating the percentage of organic C in the end.

## 13.3 PRINCIPLE

Organic carbon present in the organic matter is oxidized by chromic acid in the presence of concentrated sulfuric acid. Potassium dichromate on reaction with $H_2SO_4$ provides nascent oxygen, which combines with carbon and forms $CO_2$. The $H_2SO_4$ enables easy digestion of organic matter by rendering heat of dilution. Only certain quantity of chromic acid is used for oxidation. The excess chromic acid left unused by the organic matter is determined by back titration with 0.5*N* ferrous sulfate or ferrous ammonium sulfate using diphenylamine indicator.

## 13.4 REACTIONS

$$2\,K_2Cr_2O_7 + 8\,H_2SO_4 \longrightarrow 2\,K_2SO_4 + 2\,Cr_2(SO_4)_3 + 8H_2O + 6\,(O)$$

$$3\,C + 6\,(O) \longrightarrow 3CO_2$$

## 13.5 APPARATUS AND MATERIALS REQUIRED

- 500 mL conical flask
- Pipette
- Burette
- Measuring cylinder

## 13.6 REAGENTS REQUIRED

- 1*N* potassium dichromate ($K_2Cr_2O_7$)
- Concentrated $H_2SO_4$
- 0.5*N* ferrous sulfate (FS) or ferrous ammonium sulfate (FAS)
- Phosphoric acid (ortho phosphoric acid, 85%)
- Diphenylamine indicator

## 13.7 PROCEDURE

- Weigh exactly 0.5g of soil (passed through 0.5mm sieve) and transfer it to a 500mL conical flask.
- Add 10mL 1*N* $K_2Cr_2O_7$ and mix well by swirling the flask. Then add 20mL concentrated $H_2SO_4$ and mix by gentle rotation for one minute to ensure complete contact of the reagent with the soil.

- Run a blank simultaneously (without soil, others remaining same).
- Allow the contents to stand for 20–30 minutes. Keep the flask preferably on an asbestos sheet to avoid burning of table due to intense heat.
- Add 200mL distilled water to dilute the solution.
- Add 10mL orthophosphoric acid and add 1mL diphenylamine indicator.
- Titrate the solution with 0.5*N* ferrous ammonium sulfate.
- The color is dull green at the beginning and then shifts to a turbid blue as the titration proceeds. The end point is very sharp and shifts to a bright green color.

## 13.8 OBSERVATION AND CALCULATION

| | | |
|---|---|---|
| Weight of soil taken | = | 0.5g |
| Volume of 1*N* $K_2Cr_2O_7$ used | = | 10mL |
| Volume of 0.5*N* ferrous ammonium sulfate (FAS) used for blank titration | = | X mL |
| Volume of 0.5*N* FAS used for sample titration | = | Y mL |
| X mL FAS reduces | = | 10mL 1*N* $K_2Cr_2O_7$ |
| Therefore, Y mL of FAS reduces | = | Y/X × 10mL |
| Hence, actual quantity of 1*N* $K_2Cr_2O_7$ used for oxidation of organic matter | = | 10-(10 × Y/X) mL |
| 1mL of 1*N* $K_2Cr_2O_7$ | = | 0.003g of C |
| Therefore 10-(10 × Y/X) mL of 1*N* $K_2Cr_2O_7$ | = | [10-(10 × Y/X)] × 0.003 g of C |
| This is present in 0.5g of soil | | |
| Therefore in 100g of soil (or) % of organic carbon (OC) in the soil (uncorrected) | = | [10-(10 × Y/X)] × 0.003 × (100/0.5) |
| % of organic carbon in the soil (corrected) | = | [10-(10 x Y/X)] × 0.003 × (100/0.5) × 1.3 |
| Organic matter (surface soil) | = | OC × 1.724 |
| Organic matter (subsurface soil) | = | OC × 2.5 |

## 13.9 INTERPRETATION (FERTILITY RATING)

| Amount of OC (%) | Rating/comments |
|---|---|
| ≤ 0.5 | Low |
| 0.5–0.75 | Medium |
| 0.75 | High |

## 13.10 THINGS TO LEARN

- Why are we adding orthophosphoric acid in organic carbon estimation?
- What is the role of sulfuric acid in organic carbon estimation?
- What is the status of organic C in Indian soils?
- How will you derive the conversion factor of 1.724?
- Why a correction factor of 1.3 is used in calculating the percentage of organic C in soil?

# 14 Determination of Available Nitrogen in Soil

## *[Alkaline Permanganate/ Subbiah and Asija (1956) Method]*

## 14.1 INTRODUCTION

Nitrogen (N) is found in the arable horizon of the soil, mostly in organic material. In the soil solution, organic N is gradually transformed into ammoniacal ($NH_4^+$), nitrite ($NO_2^-$), and nitrate ($NO_3^-$) nitrogen by microbial processes. Organic-N is, in itself, of very little use to plants, as it cannot be absorbed as such. It is, therefore, necessary to estimate the different forms of mineralized or available N. The $NO_3^-$-N and $NO_2^-$-N together, hardly, exceed 1% of the total N in normal soil.

The available N in soil refers to a fraction of the total N which is converted into forms accessible to the plants. This constitutes, on an average, only 0.5–2.5% (rarely 5%) of the total N in a soil at any given time. Nitrogen is, generally, absorbed by plants as $NO_3^-$, under oxidized environment (upland conditions), and as $NH_4^+$, under reduced conditions. $NO_2^-$ is sometimes detectable, but the amount is, generally, very small vis-à-vis $NH_4^+$ or $NO_3^-$ and, hence does not warrant its determination. The sum total of $NH_4^+$ and $NO_3^-$-N is smaller than the total N that becomes available to plants, the difference being attributed to the mineralization of the organic N, during the crop growth cycle.

## 14.2 IMPORTANCE

- The chemical analysis of the soil for N is less precise when the requirement for this element needs to be forecast over a long period of time, as they vary not only with species, but with the phase of growth and season as well.
- Therefore, the chemical analysis for $NH_4^+$ and $NO_3^-$ signifies the momentary status when the sample is taken (valid for 15–20 days), and measures must be taken instantaneously.
- Nevertheless, the analysis of the extractable N content of soils, using a given extractant method, in relation to the crop response study, provides a basis for finding out its soil critical level, in order to place soils into categories,

DOI: 10.1201/9781003430100-14

on the basis of N-fertility levels, which will rationalize the use-efficiency of N-fertilizer in soils, under a given climatic condition.

## 14.3 PRINCIPLE

A given weight of soil is treated with excess of alkaline $KMnO_4$ and distilled. $KMnO_4$ is a mild oxidizing agent in an alkaline medium. The organic matter present in the soil is oxidized by the nascent oxygen, liberated by $KMnO_4$, in the presence of NaOH, and, thus, the ammonia released is distilled, and absorbed in a known volume of a standard acid, the excess of which is titrated with a standard alkali, using methyl red as the indicator. N estimated by this method is considered to be hydrolysable N or potentially available N.

## 14.4 REACTIONS

**A. Distillation**

$$2\,KMnO_4 + H_2O \xrightarrow{\text{alkaline medium}} 2\,MnO_2 + 2\,KOH + 3\,O^-$$

$$RCH.NH_2COOH + O^- \xrightarrow{\text{oxidative deamination}} R.CO.COOH + NH_3$$

$$NH_3 + H_2O \xrightarrow{\text{distillation}} NH_4OH$$

$$2NH_4OH + H_2SO_4 \xrightarrow{\text{absorption}} (NH_4)_2SO_4 + 2\,H_2O$$

**B. Titration**

$$H_2SO_4 + 2NaOH \longrightarrow Na_2SO_4 + 2H_2O$$

## 14.5 APPARATUS AND MATERIALS REQUIRED

- Kjeldahl distillation set
- Measuring cylinder
- Pipette
- Burette
- Conical flask
- Heater or burner

## 14.6 REAGENTS REQUIRED

- 0.32 percent $KMnO_4$ solution—dissolve 3.2g of $KMnO_4$ in distilled water, and make up the volume to 1 liter.
- 2.5% NaOH solution—dissolve 25g of NaOH pellets in distilled water, and make up the volume to 1 liter.
- *N*/50 $H_2SO_4$.

- *N*/50 NaOH.
- Methyl red indicator (0.15%)—dissolve 0.15g of methyl red powder in alcohol and make up the volume to 100mL.
- Glass beads and liquid paraffin or paraffin wax.

## 14.7 PROCEDURE

- Place 20g soil in a distillation flask. Add 20mL water just to moisten the soil (do not add excess water) and 1mL liquid paraffin or 1g paraffin wax.
- Add few glass beads and 100mL freshly prepared 0.32% $KMnO_4$ solution to the soil in the distillation flask.
- Pipette out 25mL *N*/50 $H_2SO_4$ in a conical flask. Add 2–3 drops of methyl red indicator, and dip the end of the delivery tube into it.
- Pour 100mL 2.5% NaOH solution into the flask and cork immediately.
- Distil the ammonia gas from the distillation flask and collect in $H_2SO_4$ solution. Continue the distillation till the evolution of ammonia ceases completely (test by bringing a moist red litmus paper near the outlet of the condenser, which will turn blue as long as ammonia is being evolved).
- Titrate the excess of $H_2SO_4$ against *N*/50 NaOH and note the volume of NaOH used. The end point is reached when the color changes from pink to yellow.

## 14.8 OBSERVATION AND CALCULATION

| | | |
|---|---|---|
| Weight of the soil sample taken | = | 20g |
| Volume of *N*/50 $H_2SO_4$ taken | = | 25mL |
| Volume of *N*/50 NaOH used (titer value) | = | X mL |
| Volume of *N*/50 $H_2SO_4$ consumed for $NH_3$ absorption | = | (25 – X) mL |
| 1mL of *N*/50 $H_2SO_4$ | = | 0.02meq / 0.28mg / 0.0014g of N |
| ∴ (25 – X) mL of *N*/50 $H_2SO_4$ | = | 0.0014 × (25 – X) g of N |
| This is present in 20g of soil | | |
| ∴ in 100g of soil/% of available N in the given soil sample | = | 0.0014 × (25 – X) × (100/20) |
| Available N in the given soil sample (ppm) | = | 0.0014 × (25 – X) × (100/20) × 10000 |
| Available N in the given soil sample (kg/ha) | = | Available N (ppm) × 2.24 |

## 14.9 INTERPRETATION (FERTILITY RATING)

| Amount of N (kg/ha) | Rating/comments |
|---|---|
| ≤ 280 | Low |
| 280–450 | Medium |
| > 450 | High |

## 14.10 THINGS TO LEARN

- Why are liquid paraffin and glass beads added during distillation?
- Can sources of organic matter influence the available N content of soils?
- What is the role of liquid paraffin/paraffin wax in the estimation of N in organic soils?
- What is the status of available N in Indian soils?
- How 1kg/ha = ppm × 2.24?

# 15 Estimation of Available Phosphorus in Soil *[Bray's Method for Acid Soils; Olsen's Method for Neutral and Alkaline Soils]*

## 15.1 INTRODUCTION

The term available phosphorus (P) refers to the inorganic form, occurring in the soil solution, which is almost exclusively "orthophosphate." This orthophosphate occurs in several forms and combinations, and only a small fraction of the total amount present may be available to plants, which is of direct relevance in assessing the P fertility level.

The phosphate concentration in solution is governed by heterogeneous equilibria in which it takes part. This situation can be represented as follows:

P adsorbed in solid phase ↔ P in the soil solution ↔ P precipitated

The phosphorus absorbed by plants from soil solution comes from the soil solution in which it exists as inorganic orthophosphate ions *viz.*, $H_2PO_4^-$, $HPO_4^{2-}$, and $PO_4^{3-}$. The most accessible ion is $H_2PO_4^-$, with the greatest activity coefficient, followed by $HPO_4^{2-}$. The quantity of P accessible to the plants is influenced by a series of soil properties. The relative abundance of these ions is, however, relatively dependent on the soil pH. For soils having a pH between 4.5 and 7.5, ions of $H_2PO_4^-$ as well as $HPO_4^{2-}$ exist in soil solution. At a pH of 7.2, $H_2PO_4^-$ and $HPO_4^{2-}$ ions have an equal activity, and, when the pH is strongly alkaline (>8.3), ions of $HPO_4^{2-}$ predominate in solution. Above pH of 9.0, the trivalent ion ($PO_4^{3-}$) becomes more important than $H_2PO_4^-$, but even at a pH of 12, the $HPO_4^{2-}$ concentration is still greater than that of $PO_4^{3-}$. Soluble phosphorus may be adsorptively retained at the surface of colloidal particles. This retention is more marked when the higher amounts of clay and sesquioxides are present. Relatively, the available inorganic P tends to accumulate in its most stable state under prevailing conditions, thus, in calcareous soils, the available inorganic P would be acid-soluble, whereas, in acid soils the adsorbed P would be more available. In flooded soils, with low oxidation potential, certain forms of P, normally considered as unavailable, for example, iron and phosphate, can be regarded as available. In most soils, the main source of orthophosphate is organic matter unless, of course, direct fertilization, with soluble phosphate, has been made.

DOI: 10.1201/9781003430100-15

## 15.2 IMPORTANCE

- The available P is considered to be a fairly good indicator/measure of the P supplying capacity of a soil.
- A knowledge of available P content of soils is important for determining its crucial limit, based on soil-test-crop-response calibration study.
- The evaluation of the soil critical limit of P would help in developing P-fertility classes for effective fertilizer recommendation schedule.
- The critical limit of soil P would help eventually in predicting crop response to the applied P.

## 15.3 ESTIMATION OF AVAILABLE PHOSPHORUS

### 15.3.1 Bray Method for Acid Soils (Bray and Kurtz, 1945)

#### 15.3.1.1 Principle

This procedure is primarily meant for soils which are moderately to strong acids (pH around 5.5 or less). It has been employed widely to give results that are highly correlated with crop response to phosphate fertilization.

The underlying principle of this method is that the soil is shaken with an extracting solution of 0.03*N* $NH_4F$ in 0.025*N* HCl, which dissolves the fractions of P that are considered to be available to plant roots. Ammonium fluoride complexes Al and Fe ions in the acid solution, with the consequent release of P held by the soil, by these trivalent ions. Thus, the combination of HCl and $NH_4F$ helps in removing easily acid-soluble forms of P, largely calcium phosphates, and a portion of the so-called aluminum and iron phosphates.

HCl results in the dissolution of more active calcium phosphate and, in addition to this, it prevents the precipitation of P (as calcium phosphate), which has been dissolved by ammonium fluoride.

In the filtered extract, P is estimated calorimetrically by adding ammonium molybdate and, thereafter, reducing the molybdenum-phosphate complex with stannous chloride in the acidic medium. The heteropoly complexes (phosphomolybdates) are formed by coordination of molybdate ions with the P as the central coordinating atom, the oxygen of the molybdate radicals being substituted for that of $PO_4$.

The heteropoly complexes, before reduction, give a yellow color hue to their water solution. In the solution of low P concentration, the molybdate is partially reduced to produce a characteristic blue color. The intensity, a measure for the concentration of P in the test solution, is read on a colorimeter.

#### 15.3.1.2 Reactions

*15.3.1.2.1 Extraction*

$$3\ NH_4F + 3\ HF + AlPO_4 \longrightarrow H_3PO_4 + (NH_4)_3AlF_6$$

$$3\ NH_4F + 3\ HF + FePO_4 \longrightarrow H_3PO_4 + (NH_4)_3FeF_6$$

The formulae $AlPO_4$ and $FePO_4$ represent various hydrated and hydroxyl phosphates of Al and Fe, including any adsorbed or precipitated surface layers on oxides of Al, aluminosilicates, and on oxides of Fe.

*15.3.1.2.2 Estimation*

$$\underset{\text{(phosphate)}}{H_3PO_4} + \underset{\text{(molybdic acid)}}{12\,H_2MoO_4} \longrightarrow \underset{\substack{\text{(phosphomolybdate)}\\\text{(yellow colored)}}}{H_3P\,(Mo_3O_{10})_4} + 12\,H_2O$$

### 15.3.1.3 Apparatus and Materials Required

- 100 mL conical flask
- 5 mL pipette
- 25 mL volumetric flask
- 50 mL measuring cylinder
- 100 mL polythene shaking bottle
- Funnel
- Whatman no. 42/44 filter paper
- Mechanical shaker
- Photoelectric colorimeter/spectrophotometer

### 15.3.1.4 Reagents Required

- 1 *N* ammonium fluoride ($NH_4F$)—dissolve 37.0g ammonium fluoride in distilled water and dilute the solution to 1 liter and store it in a polythene bottle.
- 0.5 *N* hydrochloric acid—dilute 20.2mL of concentrated HCl to a volume of 500mL with distilled water.
- Extracting solution (Bray no.1 extractant)—add 15mL of 1 *N* $NH_4F$ (37g $NH_4F$ is dissolved in 1 liter of distilled water—1 *N* $NH_4F$) and 25mL of 0.5 *N* HCl (20.8mL of concentrated HCl is diluted to 500mL with distilled water—0.5 *N* HCl) to 460mL distilled water. This gives a solution composition of 0.03 *N* $NH_4F$ in 0.025 *N* HCl.
- Dickman and Bray's reagent—dissolve 15g ammonium molybdate (AR grade), $(NH_4)_6Mo_7O_{24}.\ 4H_2O$ in 300mL of distilled water, warm to about 60°C and filter, if necessary, after cooling. Add to it 342mL of concentrated HCl and make up the volume to 1 liter. This is 1.5% solution of ammonium molybdate in HCl.
- Stannous chloride ($SnCl_2.\ 2\ H_2O$) solution—dissolve 10g of stannous chloride (AR grade) crystals in 25mL of concentrated HCl and store in a brown bottle. This is 40% $SnCl_2$ stock solution. A piece of tin metal (AR grade), if added, will keep the stock solution for longer.
- Stannous chloride working solution—dilute 0.5mL of the stock solution to 66mL with distilled water and prepare this solution just before use.
- Preparation of standard P solution—dissolve 0.439 of potassium dihydrogen phosphate (AR grade), $KH_2PO_4$, in about half a liter of distilled water. Add to it 25mL 7 *N* $H_2SO_4$ (approximately) and make up the volume to 1 liter with distilled water. This gives 100ppm stock solution of P. From this prepare a 2ppm solution by 50 times dilution of the stock solution.

### 15.3.1.5 Procedure

#### *15.3.1.5.1 Extraction*

- Weigh 5g soil and transfer it to a 100mL conical flask.
- Add 50mL extractant solution to the soil.
- Shake the contents of the flask for exactly five minutes, and filter through Whatman no. 42 filter paper. The filtrate should be clear.
- Prepare a blank in which all the reagents are added similarly, except the soil.

#### *15.3.1.5.2 Estimation (Colorimetry—Dickman and Bray, 1940)*

- Take 5mL of the soil extract as well as different concentration of P (in the range of 0 to 0.8 ppm) by pipetting out 0 (blank), 1, 2, 3, 4, 5, and 10mL of 2ppm P solution in 25mL volumetric flasks.
- Add 5mL extracting solution to each of the P standard solutions, followed by the addition of 5mL of Dickman and Bray's reagent in all the flasks (test and P standard).
- To avoid interference of fluoride, 7.5mL of 0.8 ***M*** boric acid (50g of $H_3BO_3$ per liter of distilled water) may be added, if necessary.
- Mix, thoroughly the content of the flasks with about 5mL distilled water, washing the neck of the flask down, to remove the adhering ammonium molybdate.
- Finally add 1mL working (dilute) stannous chloride solution with immediate mixing, and make upto the mark with distilled water. Once again, mix the solution thoroughly.
- Plot the transmittance readings (percent T) of the standards on the ordinate (Y axis) against the concentration in ppm of P on the abscissa (X axis) on a semi logarithmic graph paper for a straight line relationship.
- Measure the intensity of the blue color just after ten minutes, in a photoelectric colorimeter using 660nm or mμ red filter, and adjusting the meter to 100% transmittance with the blank.
- Determine the concentration of P in the soil extract from the standard curve. This is very important as the color starts fading after about 15–20 minutes of the development of color.

### 15.3.1.6 Observation and Calculation

| | | |
|---|---|---|
| Weight of the soil sample taken/used in the extraction | = | 5g |
| Volume of Bray No.1 extractant used for P estimation | = | 50mL |
| Volume of extract taken for P estimation/color development | = | 5mL |
| Final volume made up to | = | 25mL |
| Transmittance (%) of the test solution (sample) | = | T |
| Concentration of P corresponding to the % transmittance as read from the standard curve | = | X ppm |
| ∴ Available P in the given soil sample (ppm) | = | X × (50/5) × (25/5) |
| Available P in the given soil sample (kg/ha) | = | X × 50 × 2.24 |
| Available $P_2O_5$ in the given soil sample (kg/ha) | = | X x 50 x 2.24 x 2.29 |

#### 15.3.1.7 Interpretation (Fertility Rating)

| Amount of P (kg/ha) | Amount of $P_2O_5$ (kg/ha) | Rating/comments |
|---|---|---|
| ≤ 24.2 | ≤ 55.4 | Low |
| 24.2–49.7 | 57.3–113.8 | Medium |
| > 49.7 | > 113.8 | High |

### 15.3.2 Olsen's Method—Neutral and Alkaline Soils (Olsen et al., 1954)

#### 15.3.2.1 Principle

This method of extraction of soil available phosphorus is suitable for neutral and alkaline soils. The $CO_3^{2-}$ ions from $NaHCO_3$ will react with $Ca^{2+}$ and $CaCO_3$ is precipitated, thus allowing the phosphorus to come into solution. Blue color is developed by ascorbic acid. The intensity of color is measured using a spectrophotometer at 660 nm.

#### 15.3.2.2 Reactions

$$H_3PO_4 + 12\ H_2MoO_4 \longrightarrow H_3P\ (Mo_3O_{10})_4 + 12\ H_2O$$

(phosphate) (Molybdic acid) → (Phosphomolybdate) (Yellow colored)

#### 15.3.2.3 Apparatus and Materials Required

- 100 mL conical flask
- 5 mL pipette
- 25 mL volumetric flask
- 50 mL measuring cylinder
- 100 mL polythene shaking bottle
- Funnel
- Whatman no. 42/44 filter paper
- Mechanical shaker
- Photoelectric colorimeter/spectrophotometer

#### 15.3.2.4 Reagents Required

- 0.5 *M* $NaHCO_3$—dissolve 42.0g of P-free sodium bicarbonate in about 500mL hot water and dilute to 1 liter. Adjust the pH to 8.5 using dilute NaOH solution or dilute HCL.
- Activated charcoal—wash pure activated charcoal or commercially available Darco G-60 with acid to make P-free.
- Dickman and Bray's reagent—dissolve 15g of ammonium molybdate (AR grade), $(NH_4)_6Mo_7O_{24}\cdot 4H_2O$ in 300mL of distilled water, warm to about 60°C and filter, if necessary, after cooling. Add to it 342mL concentrated HCl and make up the volume to 1 liter. This is 1.5% solution of ammonium molybdate in HCl.

- Stannous chloride ($SnCl_2 . 2 H_2O$) solution—dissolve 10g of stannous chloride (AR grade) crystals in 25mL of concentrated HCl and store in a brown bottle. This is 40% $SnCl_2$ stock solution. A piece of tin metal (AR grade), if added, will keep the stock solution for longer.
- Stannous chloride working solution—dilute 0.5mL of the stock solution to 66mL with distilled water and prepare this solution just before use.
- Preparation of standard P solution—dissolve 0.439 of potassium dihydrogen phosphate (AR grade), $KH_2PO_4$, in about half a liter of distilled water. Add to it 25mL of 7 *N* $H_2SO_4$ (approximately) and make up the volume to 1 liter with distilled water. This gives 100ppm stock solution of P. From this prepare a 2ppm solution by 50 times dilution of the stock solution.

### 15.3.2.5 Procedure

- Weigh 2.5g of soil sample in a 100mL conical flask.
- Add a pinch of Darco G-60 and 50mL Olsen's extractant solution to the soil.
- Shake the contents of the flask for 30 minutes.
- Filter through Whatman no. 42 filter paper. The filtrate should be clear.
- Transfer 5mL of clear and colorless filtrate into a 25mL volumetric flask.
- Gradually add (drop by drop) 5mL Dickman and Bray's reagent.
- Shake slowly and carefully to drive out the $CO_2$ evolved.
- When frothing completely ceases, add distilled water, washing down the sides, to bring the volume to about 22mL.
- Add 1mL of freshly diluted SnC12 solution, shale a little and make the volume to 25mL.
- Read the blue color intensity at 660nm wavelength using spectrophotometer.
- Run a blank simultaneously under identical manner

### 15.3.2.6 Observation and Calculation

| | | |
|---|---|---|
| Weight of the soil sample taken/used in the extraction | = | 2.5g |
| Volume of Olsen's extractant used for P estimation | = | 50mL |
| Volume of extract taken for P estimation/color development | = | 5mL |
| Final volume made up to | = | 25mL |
| Transmittance (%) of the test solution (sample) | = | T |
| Concentration of P corresponding to the % transmittance as read from the standard curve | = | X ppm |
| ∴ Available P in the given soil sample (ppm) | = | $X \times (50/2.5) \times (25/5)$ |
| Available P in the given soil sample (kg/ha) | = | $X \times 100 \times 2.24$ |
| Available $P_2O_5$ in the given soil sample (kg/ha) | = | $X \times 100 \times 2.24 \times 2.29$ |

### 15.3.2.7 Interpretation (Fertility Rating)

| Amount of P (kg/ha) | Amount of $P_2O_5$ (kg/ha) | Rating/comments |
|---|---|---|
| ≤ 11 | ≤ 25.2 | Low |
| 11–22 | 25.2–50.4 | Medium |
| > 22 | > 50.4 | High |

## 15.4 THINGS TO LEARN

- Why is Bray's Method used to estimate the available P in acid soils?
- Why is Olsen's Method to be chosen for neutral and alkaline soils?
- What are the constraints of acid soils in terms of availability of P?
- What is the status of available P in Indian soils?
- What do you mean by P fixation?
- Mention the name of the extractants commonly used for available P estimation.

# 16 Estimation of Available Potassium in Soil *[Hanway and Heidel (1952) Method]*

## 16.1 INTRODUCTION

The total potassium (K) content of a soil varies from 0.05 to 2.5%. The total K is distributed in mineral form (lattice-K, 90–98%), fixed non-exchangeable or temporarily retrograded K (1–10%) and exchangeable plus water soluble K (1–2%). Exchangeable K represents an average of 2–15% of the sum of exchangeable bases and 1–3% of the total capacity for cation exchange. The K concentration in the soil solution is on an average 0.08–3meq/liter. Both water soluble and exchangeable K are most accessible to the plant. Available K can, thus, be separated into that immediately available, which is water soluble and exchangeable, and that potentially available or fixed. The neutral normal ammonium acetate extract contains both water soluble and exchangeable K. K extraction by this extractant is considered as a suitable index of K availability in most soils, based on crop response correlation study. Thus, the K extracted by this method is equated as the available K.

## 16.2 IMPORTANCE

- The neutral normal ammonium acetate method does not permit an evaluation of the soil reserves of potassium. Nevertheless, this method characterizes the soil well enough for the momentary state of K supply.
- In order to better know the dynamics of exchangeable K and the supply status of the plant, it is, however, necessary to make two determinations of exchangeable K by the same method at different intervals (at the beginning and during the course of the growth), and draw a curve or balance of their dynamics.
- Very approximately, it may be considered that a good supply of K is available to the plant on a loamy soil, when the exchangeable K represents >0.25% of the total K.

## 16.3 PRINCIPLE

The method is based on the principle of equilibrium of soils with an exchanging cation made of the solution of neutral normal ammonium acetate, in a given soil.

DOI: 10.1201/9781003430100-16

During the equilibrium, ammonium ions exchange with the exchangeable K ions of the soil. The K content in the equilibrium solution is estimated with a flame photometer. Since $NH_4^+$ holds highly charged layers together just as K, the release of the fixed K, in an exchangeable form, is retarded during $NH_4OAc$ extraction.

## 16.4 REACTIONS

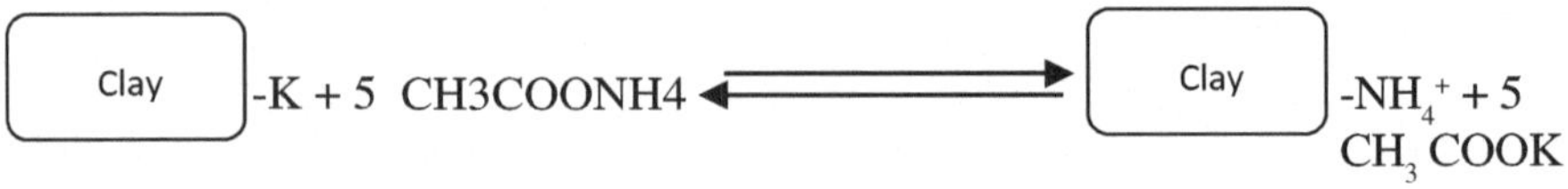


## 16.5 APPARATUS AND MATERIALS REQUIRED

- 150mL conical flask
- Funnel
- Beaker
- Pipette
- Whatman no. 1 filter paper
- Shaker
- Balance
- Flame photometer

## 16.6 REAGENTS REQUIRED

- Neutral *N* ammonium acetate ($NH_4OAc$) solution—dissolve 77.09g $NH_4OAc$ in distilled water and make up the volume to 1 liter. Adjust the solution pH to 7.0, adding acetic acid or $NH_4OH$ solution as required.
- Potassium chloride solution—dissolve 1.908 g of AR grade KCl in distilled water, and make up the volume to 1 liter. It gives 1000ppm K solution and is treated as stock solution of K.
- Standard curve for K (working K standards)—from the stock solution, take measured aliquots and dilute with $NH_4OAc$ solution to give 10–40ppm of K. After inserting the K filter and regulating the appropriate gas and air pressure, set the flame photometer reading at zero for the blank ($NH_4OAc$ solution) and at 100 for 40ppm of K solution. Draw the calibration curve by plotting the flame photometer reading against different concentrations (10, 15, 20, 25, 30, 35, 40ppm) of K. Any fluctuation in gas and air pressure does not allow steady reading in the meter and must be taken care of.

## 16.7 PROCEDURE

- Weigh 5g of soil in a 150mL conical flask.
- Add to it 25mL of neutral *N* $NH_4OAc$ solution.

- Shake the content of the conical flask on an electric shaker for five minutes and filter using Whatman no. 1 filter paper.
- Feed the filtrate into the atomizer of the flame photometer, 100 of which has been set with 40ppm K solution and note the reading.
- Locate this reading on the standard curve, which will give the K concentration in the extract.
- From this concentration measurement, the amount of K in the sample is calculated.

### 16.7.1 Precautions

- The filtrate should be clear. If some soil particles are in the filtrate, it should be refiltered to avoid the clogging of the suction capillary.
- The air pressure should be between 0.4 and 0.6kg/cm$^2$. It should not deviate too much.
- The gas inlet should be opened after opening the air inlet, and closed before shutting off the air supply.
- The flame should be soot-free and blue.
- The flame photometer should be warmed up for 10–15 minutes before using.

## 16.8 OBSERVATION AND CALCULATION

| | | |
|---|---|---|
| Weight of the soil sample taken | = | 5g |
| Volume of the neutral *N* $NH_4OAc$ (extractant) used | = | 25mL |
| Reading of the Flame Photometer for the test solution | = | R |
| Concentration of K as read from the standard curve | = | C ppm or µg/mL |
| Therefore in 25mL of solution | = | C × 25 ppm |
| This is present in 5g of soil | | |
| ∴ Available K in the given soil sample (ppm) | = | C × (25/5) ppm |
| Available K in the given soil sample (kg/ha) | = | C × 5 × 2.24 |
| Available $K_2O$ in the given soil sample (kg/ha) | = | C × 5 × 2.24 × 1.20 |

## 16.9 INTERPRETATION (FERTILITY RATING)

| Amount of K (kg/ha) | Amount of $K_2O$ (kg/ha) | Rating/comments |
|---|---|---|
| ≤ 118 | ≤ 136 | Low |
| 118–280 | 136–337.5 | Medium |
| > 280 | > 337.5 | High |

## 16.10 THINGS TO LEARN

- What are the elements that can be estimated using flame photometer?
- What are the soil types in which K availability and fixation are more?
- What is the status of available K in Indian soils?

# 17 Estimation of Available Sulfur in Soil *[Williams and Steinbergs (1959) Method]*

## 17.1 INTRODUCTION

Sulfur (S) occurs in numerous forms in soil, *viz.* sulphites, sulfates, sulphides, and in organic compounds. It is considered that the most accessible form is "sulfate" ($SO_4^{2-}$). The organic forms of S compounds become assimilable, especially following, microbiological transformation into $SO_4^{2-}$. The mineralization of organic S in a soil depends primarily upon the N:S ratio, and any $SO_4^{2-}$ formed may be fixed against extraction, particularly if much Fe or Ba is present or if the soil is very acid. Despite the fact that the plants absorb S almost exclusively as $SO_4^{2-}$, mobility of $SO_4^{2-}$ in soil may not always yield satisfactory results in accordance with the time of sampling, while assessing $SO_4^{2-}$ availability.

## 17.2 IMPORTANCE

- Development of chemical test for the estimation of available S (as $SO_4^{2-}$).is of recent interest in soil testing work, particularly in some reported S deficient areas, and for certain crops, whose requirement of this nutrient is high enough, often, exceeding that of P.
- Of several extracting procedures, suggested for available S determination, starting from distilled water for dissolving out the sulfate form to different chemical extractants *viz.*, Olsen's, Morgan's, Bray's, normal magnesium acetate, 1.0% NaCl, 0.15% $CaCl_2$ and phosphate solution.
- For estimating the plant available S in soils, heat soluble S and 0.15% $CaCl_2$ extractable S methods (Williams and Steinbergs, 1959) are usually recommended.

## 17.3 ESTIMATION OF AVAILABLE SULFUR

### 17.3.1 Heat Soluble S

#### 17.3.1.1 Introduction

Inorganic S ($SO_4^{2-}$) and a portion of organically held $SO_4^{2-}$-S are mobilized by special heat treatments and extraction with 1.0% NaCl. The $SO_4^{2-}$-S is converted into $BaSO_4$

DOI: 10.1201/9781003430100-17

suspension using $BaCl_2$ crystals and a special conditioning agent. The resulting turbidity is determined by a spectrophotometer.

#### 17.3.1.2 Apparatus and Materials Required

- 10mL and 25mL measuring cylinders
- 25mL volumetric flask
- 1mL and 5 mL pipettes
- Erlenmeyer flask
- Centrifuge tube
- Reciprocating shaker
- Water bath
- Magnetic stirring bar
- Hot air oven
- Spectrophotometer/photoelectric colorimeter

#### 17.3.1.3 Reagents Required

- 1.0% sodium chloride (NaCl)—dissolve 10.0g of NaCl in 700mL distilled water and make up to 1 liter with distilled water.
- Barium chloride dihydrate ($BaCl_2$. 2 $H_2O$) crystals (20–30 mesh)
- Conditioning agent—dissolve 75g NaCl in 275mL distilled water in a 500mL volumetric flask, stirring with magnetic stirring bar, add 30mL of concentrated HCl, 100mL of absolute ethanol and 50mL of glycerol. Rinse glycerol into flask. Continue stirring until NaCl dissolves. Remove stirring bar and make to volume with distilled water.
- Standard sulfate solution—dissolve 0.5434g of $K_2SO_4$ in 1 liter of distilled water. This contains 100mg of S per milliliter of solution. Dilute ten times to obtain 10ppm solution. Transfer 0, 5, 10, 15, and 20mL (equivalent to 0, 50, 100, 150 and 200µg of $SO_4$–S) to Erlenmeyer flasks of 150mL, bringing the volume with distilled water to 30mL.

#### 17.3.1.4 Procedure

- Weigh 5g of soil into a silica basin and add 20mL distilled water.
- Place the basin on a gently boiling water bath and evaporate to dryness.
- Then, heat it in a hot air oven at 102°C for 60 minutes.
- After cooling, transfer the soil to a 50mL centrifuge tube and extract it with 33mL of 1% NaCl.
- Draw 25mL aliquot into a silica basin.
- Evaporate to dryness with 2mL of 3% $H_2O_2$ to remove the interfering organic matter.
- Heat the basin in a hot air oven at 102°C for 60 minutes to remove the excess $H_2O_2$.
- After cooling, take up the residue in 25mL water.
- Transfer it to a centrifuge tube and centrifuge to remove the suspended matter.
- Take suitable aliquot and determine the S in it.

- Pipette out 10 or 20mL of extract in a 150mL Erlenmeyer flask and add 20 or 10mL of $H_2O_2$, thus bringing the volume to 30mL.
- Add 2.5mL of conditioning reagent and 0.2 to 0.3g of $BaCl_2$ crystals by a spatula to the standards and extracts.
- Shake the flasks for one minute each at constant rate.
- After one to three minutes, measure the turbidity in a colorimeter using a blue filter.
- Turbidity can be more accurately measured at 340nm on a spectrophotometer. The turbidity remains constraint for three to ten minutes.

## 17.3.2 $CaCl_2$ Extractable-S

### 17.3.2.1 Introduction

In this method, S is extracted with 0.15% $CaCl_2$ and the soluble sulfate is estimated turbidimetrically on a colorimeter using a blue filter or spectrophotometer at 340nm. In neutral to alkaline soils, 0.15% $CaCl_2$ may serve as a good extractant as $Ca(H_2PO_4)_2$.

### 17.3.2.2 Apparatus and Materials Required

- 150mL Erlenmeyer flask
- Volumetric flask
- Measuring cylinder
- Whatman no. 42 filter paper
- Reciprocating shaker
- Spectrophotometer/photoelectric colorimeter

### 17.3.2.3 Reagents Required

- 0.15% calcium chloride dihydrate ($CaCl_2.2H_2O$)—dissolve 1.5g of $CaCl_2.2H_2O$ in about 700mL water and make up to 1 liter with distilled water.
- 1.0% sodium chloride (NaCl)—dissolve 10.0g of NaCl in 700mL distilled water and make up to 1 liter with distilled water.
- Barium chloride dihydrate ($BaCl_2$. 2 $H_2O$) crystals (20–30 mesh).
- Conditioning agent—dissolve 75g NaCl in 275mL distilled water in a 500mL volumetric flask, stirring with magnetic stirring bar, add 30mL of concentrated HCl, 100mL absolute ethanol, and 50mL glycerol. Rinse glycerol into flask. Continue stirring until NaCl dissolves. Remove stirring bar and make to volume with distilled water.
- Standard sulfate solution—dissolve 0.5434g of $K_2SO_4$ in 1 liter of distilled water. This contains 100mg of S per milliliter of solution. Dilute ten times to obtain 10ppm solution. Transfer 0, 5, 10, 15, and 20mL (equivalent to 0, 50, 100, 150, and 200μg of $SO_4$– S) to Erlenmeyer flasks of 150mL, bringing the volume with distilled water to 30mL.

### 17.3.2.4 Procedure

- Weigh 5g of soil into a 150mL Erlenmeyer flask.
- Add 25mL of 0.15% calcium chloride to it.

- Shake it for 30 minutes on a reciprocating shaker (180 oscillations per minute).
- Filter the suspension through Whatman no. 42 filter paper.
- Analyze 10 to 20 mL of the extract turbidimetrically as described in the heat-soluble S.

## 17.4 OBSERVATION AND CALCULATION

| | | |
|---|---|---|
| Weight of the soil sample taken | = | W g |
| Volume of the aliquot used | = | X mL |
| Volume of the extract prepared | = | Y mL |
| Concentration of $SO_4$-S corresponding to the % transmittance as read from the standard curve | = | C ppm or mg/liter |
| ∴ Available sulfur ($SO_4$–S) present in the given soil sample (mg/kg) | = | $(C \times Y) / (X \times W)$ |

## 17.5 INTERPRETATION

| Amount of 0.15% $CaCl_2$-S (mg/kg) | Rating/comments |
|---|---|
| ≤ 5 | Very Low |
| 5–10 | Low |
| 10–20 | Medium |
| 20–40 | Moderately High |
| > 40 | High |

## 17.6 THINGS TO LEARN

- What is the role of conditioning reagent in the estimation of available S?
- What is the basic principle involved in turbidometry?
- What is the role of $H_2O_2$ in the estimation of available S?
- What is the status of available S in Indian soils?
- Give the various forms of S present in soil.
- How 1 ppm = 1 mg/liter?

# 18 Estimation of Available Micronutrients in Soil *[DTPA, Lindsay and Norvell (1978) Method]*

## 18.1 PRINCIPLE

DTPA (diethylene triamine penta acetic acid), a chelating agent, combines with free metal ions in solution and forms soluble complexes. Due to the reduced ion activity in the solution, desorption takes place, bringing in some more ions from solid phase. DTPA offers the most favorable combination of stability constants for the simultaneous complexing of Zn, Cu, Fe, and Mn. Since Fe and Zn deficiencies are frequently experienced in calcareous soils, the method is designed to avoid excessive dissolution of $CaCO_3$ with the release of occluded micronutrients, which are normally not available to plants. This achieved by (1) the inclusion of soluble $Ca^{2+}$, and (2) buffering the reagent at pH 7.3 with triethanolamine (TEA), which burns cleanly during flame atomization. When the extractant is added to soil, additional $Ca^{2+}$ and some $Mg^{2+}$ enter the solution. This is largely because the protonated TEA exchanges with these ions from the exchange sites and this leads to the increased ionic concentration of $Ca^{2+}$ in the solution, which in turn helps in suppressing the dissolution of $CaCO_3$. DTPA extractant has the ability to chelate Zn, Cu, Fe, and Mn in competition with $Ca^{2+}$ and $Mg^{2+}$, and unlike most other chelating agents, it applies a moderate stress to solubilize soil Fe at a pH where $CaCO_3$ is not dissolved. Suitability of this method has been proved through excellent relationships between the test values and plant utilizable nutrients under pot and field studies conducted around the world.

## 18.2 DETERMINATION OF AVAILABLE ZINC

### 18.2.1 Apparatus and Materials Required

- Atomic absorption spectrophotometer (AAS)
- Mechanical shaker

### 18.2.2 Reagents Required

1. Dilute HCl—dissolve AR grade HCl five times with double distilled water (DDW).

DOI: 10.1201/9781003430100-18

2. DTPA extractant—dissolve 1.967g of AR grade diethylene-triamine-penta acetic acid (DTPA) and 1.470g of $CaCl_2.2H_2O$ (AR grade) in about 25mL of double distilled water (DDW) by adding 13.3mL of triethanolamine (TEA), followed by 100mL more of DDW. Transfer the solution to 1 liter volumetric flask giving four to five washings. Just before making up the volume, adjust the pH to 7.3 with dilute HCl. This reagent has 0.005***M*** DTPA, 0.1***M*** TEA, and 0.01***M*** $CaCl_2.2H_2O$.
3. Standard stock solution "A" (1000ppm Zn)—weigh exactly 1.0g of pure Zn metal (AR grade) and dissolve it in minimum volume (about 10mL) of dilute HCl (1:1 with DDW) and make the volume to 1 liter.
4. Standard solution "B"—dilute 5mL of solution "A" to 100mL to get solution "B" containing 50ppm of Zn.
5. Standard working solutions—dilute 0.5, 1.0, 1.5, 2.0, 2.5, and 5.0mL portions of solution "B" to 50mL to get working standards containing 0.5, 1.0, 1.5, 2.0, 2.5, and 5.0ppm of Zn respectively. The working standards should be prepared in the medium of the extracting solution every few days, as these cannot be preserved for long.

### 18.2.3 Procedure

- Weigh 10g of soil sample into a 100mL conical flask.
- Add 20mL of DTPA extractant and shake for two hours on a mechanical shaker.
- Filter through Whatman no. 42 filter paper, discarding first few drops. For quick filtration, Whatman no. 1 filter paper can also be used if the filtrate is clear.
- Use the filtrate for Zn measurement on AAS (213.86nm).
- Feed the standard working solutions and prepare a standard curve by plotting AAS readings against Zn concentrations.
- From the standard curve, find out the concentration of Zn in the filtrate and calculate the available Zn content in soil by using the formulae given below.

## 18.3 DETERMINATION OF AVAILABLE COPPER

### 18.3.1 Apparatus and Materials Required

- Atomic absorption spectrophotometer (AAS)
- Mechanical shaker

### 18.3.2 Reagents Required

- Dilute HCl—dissolve AR grade HCl five times with double distilled water (DDW).
- DTPA extractant—dissolve 1.967g of AR grade diethylene-triamine-penta acetic acid (DTPA) and 1.470g of $CaCl_2.2H_2O$ (AR grade) in about 25mL of DDW by adding 13.3mL triethanolamine (TEA), followed by 100mL more of DDW. Transfer the solution to 1 liter volumetric flask giving four to five

washings. Just before making up the volume, adjust the pH to 7.3 with dilute HCl. This reagent has 0.005*M* DTPA, 0.1*M* TEA, and 0.01*M* $CaCl_2.2H_2O$.

- Standard stock solution "A" (1000ppm Cu)—weigh exactly 1.0g of AR grade copper metal wire and dissolve it in 50mL of diluted $HNO_3$ (1:1 with DDW) and finally make the volume to one liter. This solution contains 1000ppm of Cu.
- Standard solution "B"—dilute 5mL of solution "A" to 100mL to get solution "B" containing 50ppm of Cu.
- Standard working solutions—dilute 0.25, 0.5, 1.0, 1.5, 2.0, and 2.5mL portions of solution "B" to 50mL to get working standards containing 0.25, 0.5, 1.0, 1.5, 2.0, and 2.5ppm of Cu respectively. The working standards should be prepared in the medium of the extracting solution every few days, as these cannot be preserved for long.

### 18.3.3 Procedure

- Weigh 10g of soil sample into a 100mL conical flask.
- Add 20mL of DTPA extractant and shake for two hours on a mechanical shaker.
- Filter through Whatman no. 42 filter paper, discarding first few drops. For quick filtration, Whatman no. 1 filter paper can also be used if the filtrate is clear.
- Use the filtrate for Cu measurement on AAS (324.75nm).
- Feed the standard working solutions and prepare a standard curve by plotting AAS readings against Cu concentrations.
- From the standard curve, find out the concentration of Zn in the filtrate and calculate the available Cu content in soil.

## 18.4 DETERMINATION OF AVAILABLE IRON

### 18.4.1 Apparatus and Materials Required

- Atomic absorption spectrophotometer (AAS)
- Mechanical shaker

### 18.4.2 Reagents Required

- Dilute HCl—dissolve AR grade HCl five times with double distilled water (DDW).
- DTPA extractant—dissolve 1.967g of AR grade diethylene-triamine-penta acetic acid (DTPA) and 1.470g of $CaCl_2.2H_2O$ (AR grade) in about 25mL DDW by adding 13.3mL of triethanolamine (TEA), followed by 100mL more of DDW. Transfer the solution to 1 liter volumetric flask giving four to five washings. Just before making up the volume, adjust the pH to 7.3 with dilute HCl. This reagent has 0.005*M* DTPA, 0.1*M* TEA, and 0.01 *M* $CaCl_2.2H_2O$.

- Standard stock solution "A" (1000ppm Cu)—weigh exactly 1.0g of AR grade Fe metal and dissolve it in 50mL of diluted $HNO_3$ (1:1 with DDW) and finally make the volume to 1 liter. This solution contains 1000ppm of Fe.
- Standard solution "B"—dilute 50mL of solution "A" to 500mL to get solution "B" containing 100ppm of Fe.
- Standard working solutions—dilute 2.0, 4.0, 6.0, 10.0, and 20.0mL portions of solution "B" to 50mL to get working standards containing 1.0, 2.0, 3.0, 5.0, and 10.0 ppm of Fe respectively. The working standards should be prepared in the medium of the extracting solution every few days, as these cannot be preserved for long.

### 18.4.3 Procedure

- Weigh 10g of soil sample into a 100mL conical flask.
- Add 20mL of DTPA extractant and shake for two hours on a mechanical shaker.
- Filter through Whatman no. 42 filter paper, discarding first few drops. For quick filtration, Whatman no. 1 filter paper can also be used if the filtrate is clear.
- Use the filtrate for Fe measurement on AAS (248.33nm).
- Feed the standard working solutions and prepare a standard curve by plotting AAS readings against Fe concentrations.
- From the standard curve, find out the concentration of Fe in the filtrate and calculate the available Fe content in soil.

## 18.5 DETERMINATION OF AVAILABLE MANGANESE

### 18.5.1 Apparatus and Materials Required

- Atomic absorption spectrophotometer (AAS)
- Mechanical shaker

### 18.5.2 Reagents Required

- Dilute HCl—dissolve AR grade HCl five times with double distilled water (DDW).
- DTPA extractant—dissolve 1.967g of AR grade diethylene-triamine-penta acetic acid (DTPA) and 1.470g of $CaCl_2.2H_2O$ (AR grade) in about 25mL DDW by adding 13.3mL triethanolamine (TEA), followed by 100mL more of DDW. Transfer the solution to a 1 liter volumetric flask giving four to five washings. Just before making up the volume, adjust the pH to 7.3 with dilute HCl. This reagent has 0.005 ***M*** DTPA, 0.1 ***M*** TEA, and 0.01 ***M*** $CaCl_2.2H_2O$.
- Standard stock solution "A" (1000ppm Cu)—weigh exactly 1.583g of AR grade $MnO_2$ or 1.0g of pure Mn metal and dissolve it in 50mL of diluted $HNO_3$ (1:1 with DDW) and finally make the volume to 1 liter. This solution contains 1000ppm of Mn.

- Standard solution "B"—dilute 25mL of solution "A" to 250mL to get solution "B" containing 100ppm of Mn.
- Standard working solutions—dilute 0.5, 1.0, 2.0, 2.5, and 5.0mL portions of solution "B" to 100mL to get working standards containing 0.5, 1.0, 2.0, 2.5, and 5.0 ppm of Fe respectively. The working standards should be prepared in the medium of the extracting solution every few days, as these cannot be preserved for long.

### 18.5.3 Procedure

- Weigh 10g of soil sample into a 100mL conical flask.
- Add 20mL of DTPA extractant and shake for two hours on a mechanical shaker.
- Filter through Whatman no. 42 filter paper, discarding first few drops. For quick filtration, Whatman no. 1 filter paper can also be used if the filtrate is clear.
- Use the filtrate for Mn measurement on AAS (279.48nm).
- Feed the standard working solutions and prepare a standard curve by plotting AAS readings against Mn concentrations.
- From the standard curve, find out the concentration of Mn in the filtrate and calculate the available Mn content in soil.

## 18.6 OBSERVATION AND CALCULATION

| | | |
|---|---|---|
| Weight of the soil sample taken | = | W g (10g) |
| Volume of the extractant used | = | Y mL (20g) |
| Concentration of Zn/Cu/Fe/Mn in aliquot as read from X-axis of standard curve against the sample reading | = | C ppm |
| ∴ Available (DTPA-extractable) (Zn/Cu/Fe/Mn) in soil (mg/kg or ppm) | = | $C \times (Y/W)$ |
| | = | $C \times (20 / 10)$ |
| | = | $C \times 2$ |

## 18.7 THINGS TO LEARN

- What is the role of DTPA in the estimation of available micronutrients?
- What is the basic principle involved in atomic absorption spectrophotometry?
- What is the difference between AAS and flame emission spectrophotometry?
- Give the critical limit/level of various micronutrients in soils?
- What is soil-extractant ratio and shaking duration to be maintained?

# 19 Determination of Water-Soluble Carbonate and Bicarbonate in Soil *(Titrimetric Method)*

## 19.1 INTRODUCTION

The term soluble salts, as applied to soils, refers to inorganic soils constituents that are appreciably soluble in water. The determination of soluble salts, accumulating in soils, is important, particularly in areas where brackish water (water of poor quality) is frequently used for irrigation purposes.

In extreme cases, the salinity of soils may be a limiting factor in crop production. The soluble salts that are of major concern, particularly in salt-affected soils, include $CO_3^{2-}$, $HCO_3^-$, $Cl^-$, $SO_4^{2-}$, $NO_3^-$, $Ca^{2+}$, $Mg^{2+}$, $K^+$, and $Na^+$. Of these, anions like $CO_3^{2-}$ and $HCO_3^-$ are of greater importance.

## 19.2 IMPORTANCE

- Estimation of soluble salts is necessary to examine soils and/or water for saline constituents for reclamation measures and for irrigation purposes.
- Knowledge of soluble salts is also important to establish limits for essential elements in particular, which can indicate deficiencies and physiological disturbance.

## 19.3 PRINCIPLE

The estimation is based on simple acidimetry in the presence of phenolphthalein for $CO_3^{2-}$ ($pH \geq 8.5$), and, then, in the presence of methyl, orange for $HCO_3^-$ ($pH < 6.0$). When a mixture of carbonate and bicarbonate is titrated with standard $H_2SO_{4,}$ the phenolphthalein gives color so long as $CO_3^{2-}$ remains. It will be discharged as soon as all the $CO_3^{2-}$ is converted into $HCO_3^-$, $HCO_3^-$ can be titrated to neutrality with methyl orange as indicator. In the titration with methyl orange, the original $HCO_3^-$, together with the $HCO_3^-$ formed from the titration of $CO_3^{2-}$, is neutralized by the standard acid.

DOI: 10.1201/9781003430100-19

## 19.4 REACTIONS

$$Na_2CO_3 + H_2SO_4 \longrightarrow NaHSO_4 + NaHCO_3$$

$$2NaHCO_3 + H_2SO_4 \longrightarrow Na_2SO_4 + 2CO_2 + 2\,H_2O$$

## 19.5 APPARATUS AND MATERIALS REQUIRED

- Conical flask
- Porcelain dish
- Glass rod
- Burette
- Pipette
- Shaking machine

## 19.6 REAGENTS REQUIRED

- Phenolphthalein indicator (0.25% solution in 60% ethyl alcohol)
- Methyl orange indicator (0.25% solution in 95% ethyl alcohol)
- Standard $H_2SO_4$ (0.01*N*)

## 19.7 PROCEDURE

### 19.7.1 Extraction

- Weigh 40g of soil in a conical flask.
- Add 200mL of distilled water, and place the flask in a shaking machine for one hour for equilibration.
- Filter the suspension and the filtrate is used for different estimations.

### 19.7.2 Volumetric Analysis

#### 19.7.2.1 Carbonates

- Take 5mL of the extract or 5mL of filtered water sample (containing not more than 1meq of $CO_3^{2-}$ + $HCO_3^-$) in a porcelain dish.
- Add two to three drops of phenolphthalein indicator.
- Titrate the aliquot with 0.01*N* $H_2SO_4$ until the pink color just disappears (phenolphthalein end point).
- This end point corresponds to the neutralization of the carbonate to the bicarbonate stage. Note the titer value.

#### 19.7.2.1 Bicarbonates

- Add one to two drops of methyl orange indicator to the colorless solution from the titration of $CO_3^{2-}$ (or to the original extract/water sample if there was no color with phenolphthalein).

- Titrate it again with 0.01*N* $H_2SO_4$. Continue the titration with brisk stirring until the indicator turns red, denoting the complete neutralization of the bicarbonate present.
- Note the titer value and calculate the amount of carbonate and bicarbonate present in the given sample.

## 19.8 OBSERVATION AND CALCULATION

| | | |
|---|---|---|
| Weight of the soil sample taken | = | 40.0g |
| Volume of water added | = | 200mL |
| Volume of aliquot (from soil extract / water sample) taken for analysis | = | V mL |
| Volume of *N*/10 $H_2SO_4$ used in first titration (with phenolphthalein) | = | X mL |
| Total volume of *N*/10 $H_2SO_4$ used in successive titrations (with phenolphthalein + methyl orange) | = | Y mL |
| 1 mL of *N*/10 $H_2SO_4$/0.01meq of *N*/10 $H_2SO_4$ | = | 0.00030g of $CO_3^{2-}$ |
| 1 mL of *N*/10 $H_2SO_4$/0.01meq of *N*/10 $H_2SO_4$ | = | 0.00061 g of $HCO_3^-$ |
| meq of *N*/10 $H_2SO_4$ used in the first titration | = | $0.01 \times X$ |
| meq of N/10 H2SO4 used (in total) in the successive titration | = | $0.01 \times Y$ |
| This is present in V mL of the extract | | |
| ∴ meq of $CO_3^{2-}$ per liter of extract or water sample | = | $0.01 \times X \times (1000/V)$ |
| ∴ meq of $HCO_3^-$ per liter of extract or water sample | = | $0.01 \times (Y-X) \times (1000/V)$ |
| ∴ meq of $CO_3^{2-}$ per 100 g of soil | = | $0.01 \times X \times (200/V) \times (100/40)$ |
| ∴ meq of $HCO_3^-$ per 100 g of soil | = | $0.01 \times (Y-X) \times (1000/V) \times (100/40)$ |
| ∴ mg of $CO_3^{2-}$ per 100 g of soil | = | $0.01 \times X \times (200/V) \times (100/40) \times 30$ |
| ∴ mg of $HCO_3^-$ per 100 g of soil | = | $0.01 \times (Y-X) \times (1000/V) \times (100/40) \times 61$ |

## 19.9 INTERPRETATION

| Soluble salts (%) in soil in 0–20 cm layer | | |
|---|---|---|
| $CO_3^{2-}$ | $HCO_3^-$ | Comments |
| Nil | < 0.06 | Non-salinized |
| < 0.005 | – | Weakly salinized |
| 0.005–0.010 | – | Average salinized |
| 0.011–0.030 | – | Strongly salinized |

## 19.10 THINGS TO LEARN

- What is the end point and indicators used for the determination of $CO_3^{2-}$ and $HCO_3^-$?

- What is the role of $CO_3^{2-}$ and $HCO_3^-$ in salinization?
- What is the principle involved in determination of $CO_3^{2-}$ and $HCO_3^-$?
- How will you prepare phenolphthalein and methyl orange indicator?
- What is the reaction involved in the estimation of $CO_3^{2-}$ and $HCO_3^-$?

# 20 Determination of Water-Soluble Chloride in Soil *(Argentimetric Method)*

## 20.1 INTRODUCTION

The determination of soluble salts especially chloride, accumulating in soils, is important, particularly in areas where brackish water (water of poor quality) is frequently used for irrigation purposes. In extreme cases, chloride toxicity and salinity of soil may be a limiting factor in crop production.

## 20.2 IMPORTANCE

- Estimation of soluble salts especially chloride is necessary to examine soils and/or water for reclamation measures and to identify the suitability of water for irrigation purposes.
- Knowledge of chloride level is also important to establish limits for essential elements in particular, which can indicate deficiencies and physiological disturbance and especially for the determination of excess chloride, which is just as detrimental as a deficiency.

## 20.3 PRINCIPLE

The best-known reaction for chloride ($Cl^-$) determination is based on the formation of nearly insoluble silver salts, when titrated with standard silver nitrate solution using $K_2CrO_4$ as indicator. First of all, the most stable salt (AgCl) is formed and then the excessive $AgNO_3$ reacts with $K_2CrO_4$, forming a reddish-brown precipitate of $AgCrO_4$, which indicates the end point of the reaction.

## 20.4 REACTIONS

$$Cl^- + Ag \rightleftharpoons AgCl \text{ (whiter spongy precipitate)}$$

$$NaCl + AgNO_3 \rightleftharpoons AgCl + NaNO_3$$

$$K_2CrO_4 + 2\,AgNO_3 \rightleftharpoons AgCr_2O_4 \text{ (reddish brown ppt)} + 2\,KNO_3$$

DOI: 10.1201/9781003430100-20

## 20.5 APPARATUS AND MATERIALS REQUIRED

- Conical flask
- Volumetric flask
- Porcelain dish
- Burette
- Pipette
- Glass rod
- Balance
- Shaking machine

## 20.6 REAGENTS REQUIRED

- Potassium chromate indicator—5% aqueous solution of pure $K_2CrO_4$
- 0.02*N* $AgNO_3$ solution—dissolve 3.40g of $AgNO_3$ (AR grade) in double distilled water and make up to 1 liter. Standardize this solution against a standard NaCl solution, and store in amber colored bottle away from light.

## 20.7 PROCEDURE

- Take 50mL of aliquot from the same soil-water extract as that used in $CO_3^{2-}$ and $HCO_3^-$ estimation or 5mL of the filtered water sample (the one after $CO_3^{2-}$ and $HCO_3^-$ titration can also be used).
- Add to it five to six drops of $K_2CrO_4$ indicator, and titrate the solution with 0.02*N* $AgNO_3$ solution (with stirring) till the first reddish-brown tinge appears. The volume of $AgNO_3$ (titer value) required refers to the amount of chloride present in the sample.

## 20.8 OBSERVATION AND CALCULATION

| | | |
|---|---|---|
| Weight of the soil sample taken | = | 40.0g |
| Volume of water added | = | 200mL |
| Volume of aliquot (from soil extract / water sample) taken for analysis | = | V mL |
| Volume of 0.02*N* $AgNO_3$ solution used in titration (titer value) | = | X mL |
| meq of 0.02*N* $AgNO_3$ used in the titration | = | $0.02 \times X$ |
| This is present in V mL of the extract | | |
| ∴ meq of $Cl^-$ per liter of extract or water sample | = | $0.02 \times X \times (1000/V)$ |
| ∴ meq of $Cl^-$ per 100g of soil | = | $0.02 \times X \times (200/V) \times (100/40)$ |
| ∴ mg of $Cl^-$ per 100g of soil | = | $0.02 \times X \times (200/V) \times (100/40) \times 35.5$ |

## 20.9 INTERPRETATION

| **Amount of $Cl^-$ (%) in soil in 0–20 cm layer** | **Comments** |
|---|---|
| < 0.020 | Non-salinized |

| Amount of $Cl^-$ (%) in soil in 0–20 cm layer | Comments |
|---|---|
| 0.020–0.050 | Weakly salinized |
| 0.051–0.120 | Average salinized |
| 0.121–0.200 | Strongly salinized |

## 20.10 THINGS TO LEARN

- What is argentometry?
- What is the basic principle involved in the estimation of chloride?
- What is the end point and indicator used in the estimation of chloride?
- How will you prepare potassium chromate indicator?
- What is the role of chloride in salinization?

# 21 Determination of Water-Soluble Calcium and Magnesium in Soil *(Complexometric Titration Method/Versenate Method)*

## 21.1 INTRODUCTION

Calcium (Ca) and magnesium (Mg) are the two most abundant alkaline earth cations in soils. They both occur as soluble ions in soil solution, ions in the adsorbed state (exchangeable) and non-exchangeable ions in primary and secondary minerals and in organic materials.

Knowledge of the concentration and nature of the water-soluble constituents of soils especially cations like Ca and Mg is desirable in many circumstances, such as salinity and nutrient availability appraisals, exchangeable cation determinations and others. The water-soluble Ca and Mg contents in soils ranged from 2 to 200 ppm.

The most widely used method for the determination of Ca and Ca+Mg is by complexometric titration, involving ethylene diamine tetra-acetic acid (EDTA), first introduced by Schwartzenbach et al. (1946).

## 21.2 PRINCIPLE

The most widely used salt EDTA is the "disodium salt" with the formula of $Na_2H_2Y.2H_2O$, where, Y is the tetravalent anion of EDTA. EDTA forms stable complexes with various polyvalent cations at different pH level. Calcium and magnesium are complexed by EDTA in the order of calcium first and magnesium afterwards at pH levels of 12 and 10 respectively. Hence, calcium is first estimated by titration with EDTA using murexide indicator in the presence of sodium hydroxide at pH 12. Then magnesium is estimated together with calcium using Eriochrome black-T indicator at pH 10 in the presence of ammonium chloride-ammonium hydroxide buffer solution. Sn, Cu, Zn, Fe, Mn may interfere in the determination of calcium and magnesium and such interference is prevented by the use of 2% NaOH solution.

## 21.3 APPARATUS AND MATERIALS REQUIRED

- Porcelain basin
- Burette

DOI: 10.1201/9781003430100-21

- Conical flask
- Pipette
- Volumetric flask
- Beaker
- Glass rod
- Shaking machine
- Water bath

## 21.4 REAGENTS REQUIRED

- 0.02 *N* EDTA
- 10% sodium hydroxide
- Ammonium chloride–ammonium hydroxide buffer solution—dissolve 67.5g of ammonium chloride in 400mL distilled water, to this, add 570mL concentrated ammonium hydroxide and dilute to 1 liter
- Murexide indicator
- Eriochrome black-T indicator
- Red litmus paper

## 21.5 PROCEDURE

### 21.5.1 Preparation of Soil Extract

- Weigh 40g of soil in a conical flask.
- Add 200mL of distilled water, and place the flask in a shaking machine for one hour for equilibration.
- Filter the suspension and the filtrate is used for the estimations.

### 21.5.2 Pretreatment of Soil Extract

- Remove excess ammonium acetate and organic matter from the soil extract by complete evaporation of the aliquot, followed by treatment with aqua regia (concentrated HCl and concentrated $HNO_3$ in the ratio of 3:1) and a second evaporation to dryness (very dark colored soil extracts may require additional treatment with aqua regia).
- Dissolve the residue in a quantity of water equal to the original volume of the aliquot taken for the treatment.

### 21.5.3 Determination of Calcium Alone

- Pipette out 10mL of soil water extract into a porcelain basin.
- Add 10% sodium hydroxide solution drop by drop to neutralize the acidity (red litmus turns blue) and another 5mL excess to maintain the pH at 12.
- Add a pinch of murexide indicator and titrate with 0.02 *N* EDTA till the color changes from pinkish red to purple or violet.

### 21.5.4 Determination of Calcium + Magnesium

- Pipette out 10mL of soil water extract into a porcelain basin.
- Adjust to pH 10 by adding 15mL of ammonium chloride–ammonium hydroxide buffer (red litmus turns blue).
- Add 2–3 drops of Eriochrome black-T indicator solution and titrate with 0.02 *N* EDTA till the color changes from purple red to sky blue.

## 21.6 OBSERVATION AND CALCULATION

| | | |
|---|---|---|
| Weight of the soil sample taken | = | 40g |
| Volume of water added | = | 200mL |
| Volume of aliquot (from soil extract/water sample) taken for analysis | = | V mL |
| ***Calcium*** | | |
| Volume of 0.02 *N* EDTA used for calcium alone (titer-value) | = | X mL |
| 1mL of 0.02 *N* EDTA | = | 0.0004g of Ca |
| ∴ X mL of 0.02 *N* EDTA | = | X × 0.0004g of Ca |
| This is present in V mL of aliquot | | |
| ∴ in $10^6$mL | = | X × 0.0004 × ($10^6$/V) ppm |
| In terms of meq/liter | = | X × 0.0004 × (1000/V) × (1000/20) |
| ***Magnesium*** | | |
| Volume of 0.02 *N* EDTA used for calcium + magnesium | = | Y mL |
| Volume of 0.0 2*N* EDTA used for magnesium alone | = | (Y-X) mL |
| 1 mL of 0.02 *N* EDTA | = | 0.00024g of Mg |
| ∴ (Y-X) mL of 0.02 *N* EDTA | = | (Y-X) × 0.00024g of Mg |
| This is present in V mL of aliquot | | |
| ∴ in $10^6$ mL | = | (Y-X) × 0.00024 × ($10^6$/V) ppm |
| In terms of meq/liter | = | (Y-X) × 0.00024 × (1000/V) × (1000/12) |

## 21.7 THINGS TO LEARN

- How 1mL of 0.02 *N* EDTA = 0.00024g of Mg and 0.0004g of Ca?
- Give the indicators used and end point in the determination of Ca and Mg.
- What is complexometric titration? How does it help in the determination of Ca and Mg?

# 22 Determination of Water-Soluble Sodium and Potassium in Soil *(Flame Photometric Method)*

## 22.1 PRINCIPLE

Flame photometry is an essential tool for the determination of sodium (Na) and potassium (K), for which the chemical methods are time-consuming and tedious. The solution is sprayed as a fine mist into a non-luminous flame and the emitted light, after passing through a light filter or prism system so as to exclude light of the wavelengths other than the characteristic of the emission from the element being measured, is allowed to fall on a photocell. The current is generated by the photocell and will obviously vary with the concentration of the element and compared with that generated when a number of standard solutions are examined. When the sample, in liquid form, is atomized into a hot flame, it excites atoms in the sample and causes them to emit radiations at various wavelengths. Sodium emits a bright yellow color (489nm) and the potassium emits a yellow color (404nm), when excited in the flame. The intensity of emission is proportional to the concentration of the ions in the sample.

## 22.2 APPARATUS AND MATERIALS REQUIRED

- 100mL and 1000mL volumetric flask
- 100mL beaker
- Flame photometer

## 22.3 REAGENTS REQUIRED

- Standard curve for Na—dissolve 0.254 g of AR grade NaCl in distilled water, and make up to 1 liter. This gives 100ppm of Na solution. From this, prepare a series of concentrations of Na (10, 20, 30, 40, 50, 60, 70, 80, 90ppm) by transferring 10, 20, 30, 40, 50, 60, 70, 80, 90mL of 100ppm solution into a 100mL volumetric flask. After inserting the Na filter and regulating the appropriate gas and air pressure, set the flame photometer reading at zero for the blank (distilled water) and at 100 for 100ppm of Na solution. Draw the calibration curve by plotting the flame photometer reading against different concentrations (10, 20, 30, 40, 50, 60, 70, 80, 90, 100ppm of Na).

DOI: 10.1201/9781003430100-22

- Standard curve for K—dissolve 1.908g of AR grade KCl in distilled water, and make up the volume to 1 liter. It gives 1000ppm K solution and is treated as stock solution of K. From the stock solution, take measured aliquots and dilute with $NH_4OAc$ solution to give 10 to 100ppm of K. After inserting the K filter and regulating the appropriate gas and air pressure, set the flame photometer reading at zero for the blank (distilled water) and at 100 for 100ppm of K solution. Draw the calibration curve by plotting the flame photometer reading against different concentrations (10, 20, 30, 40, 50, 60, 70, 80, 90, 100ppm of K).

## 22.4 PROCEDURE

- Weigh 40g of soil in a conical flask.
- Add 200mL distilled water, and place the flask in a shaking machine for one hour for equilibration.
- Filter the suspension and the filtrate is used for the estimation of Na and K.
- Take a known amount of aliquot in a volumetric flask and make up the volume with distilled water.
- Feed the sample in the flame photometer and note the meter readings.
- From the standard curve, find out the concentrations of Na and K in the given sample.

## 22.5 OBSERVATION AND CALCULATION

| | | |
|---|---|---|
| Weight of the soil sample taken | = | 40.0g |
| Volume of water added | = | 200mL |
| Volume of aliquot (from soil extract/water sample) taken for analysis | = | X mL |
| Volume made up to | = | V mL |
| Concentration of Na as read from the standard curve of Na | = | $C_{Na}$ ppm |
| Concentration of K as read from the standard curve of K | = | $C_K$ ppm |
| ∴ Amount of water-soluble Na in the given soil sample (ppm) | = | $C_{Na} \times (200/40) \times (V/X)$ |
| ∴ Amount of water-soluble K in the given soil sample (ppm) | = | $C_k \times (200/40) \times (V/X)$ |

## 22.6 THINGS TO LEARN

- What is the principle underlying in the estimation of Na and K?
- What is flame photometry?
- How will you prepare the standard curve for Na?
- How will you prepare the standard curve for K?

# 23 Determination of Cation Exchange Capacity of Soil

## 23.1 PRINCIPLE

The cation exchange capacity of the soil involves the measurement of the total quantity of negative charges per unit weight of the soil. It is determined by leaching a known amount of soil by neutral salt solution (usually neutral normal ammonium acetate) and then determining the quantity of cations ($NH_4^+$) adsorbed by the soil by distillation.

## 23.2 APPARATUS AND MATERIALS REQUIRED

- 250mL beaker
- Whatman no. 3 filter paper
- 250mL volumetric flask
- Watch glass
- Distillation set
- 500mL ice tumbler

## 23.3 REAGENTS REQUIRED

- Neutral normal ammonium acetate
- Solid ammonium chloride
- 0.1 *N* KOH
- Red litmus paper
- 60% alcohol
- 0.1 *N* $H_2SO_4$
- Methyl red indicator
- 0.1 *N* $AgNO_3$
- 40% NaOH

## 23.4 PROCEDURE

- Put 10g of air-dry soil in a 250mL beaker and add 50mL of neutral normal ammonium acetate solution.

DOI: 10.1201/9781003430100-23

- Stir the contents well and keep it overnight, covering with a watch glass.
- Transfer the soil to the filter paper (Whatman no. 3) and leach it with 25–30mL portions of ammonium acetate solution 6–8 times, collecting the filtrate in a beaker. (Transfer the filtrate to a 250mL volumetric flask and make up the volume to 250mL. The same shall be preserved for the estimation of individual exchangeable cations.)
- Add a pinch of solid ammonium chloride to the soil in the filter paper and leach it with 60% alcohol.
- Transfer the soil in the beaker completely to the filter paper by means of 60% alcohol and wash it with alcohol until the filtrate runs free of chloride.
- Test with 0.1*N* $AgNO_3$ indicates the presence or absence of chloride. (Collect the filtrate in a test tube and add two or three drops of silver nitrate solution. If there is no precipitate formation, it is an indication of free of chloride.)
- Remove the filter paper along with the soil cautiously and place it in a distillation flask. Add about 500mL distilled water.
- Pipette out 25mL of 0.1*N* $H_2SO_4$ into a 500mL ice tumbler and add one or two drops of methyl red indicator.
- Place the ice tumbler below the delivery end and ensure that the delivery end is completely immersed in 0.1*N* $H_2SO_4$.
- Add 10mL of 40% NaOH to the flask containing the soil.
- Stopper the flask immediately and start distilling the contents.
- Collect the ammonia liberated in the acid taken in the ice tumbler. Test it is free of ammonia by using red litmus paper test (keep the moist red litmus paper at the delivery end after washing it with distilled water—if the red litmus paper does not turn into blue color, it is an indication that it is free of ammonia).
- Wash the end of the delivery tube with distilled water into the same ice tumbler when the distillate runs free of ammonia.
- Remove the ice tumbler and titrate the excess acid against 0.1*N* KOH.
- The end point is the appearance of light or straw yellow color.

## 23.5 OBSERVATION AND CALCULATION

| | | |
|---|---|---|
| Weight of the soil taken | = | 10 g |
| Volume of 0.1*N* $H_2SO_4$ taken | = | V mL |
| Volume of 0.1*N* KOH consumed for titration | = | $V_1$ mL |
| Actual volume of 0.1*N* $H_2SO_4$ consumed | = | $(V - V_1)$ mL |
| 1 mL of 0.1*N* $H_2SO_4$ | = | 0.1 meq of any ion |
| ∴ cation exchange capacity of soil | = | $(V - V_1) \times 0.1 \times (100/10)$ meq / 100 g of soil or C.mol($p^+$) / kg of soil |

## 23.6 THINGS TO LEARN

- What is the purpose of adding neutral normal ammonium acetate?
- What is the role of solid ammonium nitrate?
- Why do you wash with 60% alcohol?
- Name the latest unit of CEC.
- What is the principle involved in the determination of CEC of soil?

# 24 Determination of Exchangeable Potassium in Soil

## 24.1 PRINCIPLE

The method is based on the principle that the soil is first saturated with neutral normal ammonium acetate solution and the excess ammonium ions are removed by washing with alcohol. The displaced cations including potassium ions are collected as the filtrate in a volumetric flask during the estimation of cation exchange capacity of the soil.

The potassium present in the filtrate is then estimated using a flame photometer to find the exchangeable potassium present in the soil.

## 24.2 APPARATUS AND MATERIALS REQUIRED

- 250mL beaker
- Whatman no. 3 filter paper
- 250mL volumetric flask
- Watch glass
- 25mL volumetric flask
- Flame photometer

## 24.3 REAGENTS REQUIRED

- Neutral normal ammonium acetate.
- Standard curve for K—dissolve 1.908g of AR grade KCl in distilled water, and make up the volume to 1 liter. It gives 1000ppm K solution and is treated as stock solution of K. From the stock solution, take measured aliquots and dilute with $NH_4OAc$ solution to give 10 to 100ppm of K. After inserting the K filter and regulating the appropriate gas and air pressure, set the flame photometer reading at zero for the blank (distilled water) and at 100 for 100ppm of K solution. Draw the calibration curve by plotting the flame photometer reading against different concentrations (10, 20, 30, 40, 50, 60, 70, 80, 90, 100ppm of K).

DOI: 10.1201/9781003430100-24

## 24.4 PROCEDURE

- Put 10g of air-dry soil in a 250mL beaker and add 50mL of neutral normal ammonium acetate solution.
- Stir the contents well and keep it overnight, covering with a watch glass.
- Transfer the soil to the filter paper (Whatman no. 3) and leach it with 25–30mL portions of ammonium acetate solution 6–8 times, collecting the filtrate in a beaker.
- Transfer the filtrate to a 250mL volumetric flask and make up the volume to 250mL.
- Feed the filtrate into the flame photometer for finding the concentration of potassium present in the filtrate and the exchangeable potassium in the soil is calculated accordingly.

## 24.5 OBSERVATION AND CALCULATION

| | | |
|---|---|---|
| Weight of the soil taken | = | 10g |
| Volume of the filtrate prepared | = | 250mL |
| Concentration of potassium in the filtrate as derived from the standard graph for K | = | X ppm |
| ∴ Amount of exchangeable potassium in the given soil | = | (X / 1000) × 250 × (100 / 10) × (1/39) meq / 100g of soil or C.mol($p^+$) / kg of soil |

## 24.6 THINGS TO LEARN

- What is the purpose of adding neutral normal ammonium acetate?
- What are precautions to be followed while using the flame photometer?
- How to convert ppm into meq/100 g of soil or C.mol($p^+$)/kg of soil?

# 25 Determination of Exchangeable Sodium in Soil

## 25.1 PRINCIPLE

The method is based on the principle that the soil is first saturated with neutral normal ammonium acetate solution and the excess ammonium ions are removed by washing with alcohol. The displaced cations including sodium are collected as the filtrate in a volumetric flask during the estimation of cation exchange capacity of the soil.

The sodium present in the filtrate is then estimated by using a flame photometer to find the exchangeable sodium present in the soil.

## 25.2 APPARATUS AND MATERIALS REQUIRED

- 250mL beaker
- Whatman no. 3 filter paper
- 250mL volumetric flask
- Watch glass
- 25mL volumetric flask
- Flame photometer

## 25.3 REAGENTS REQUIRED

- Neutral normal ammonium acetate.
- Standard curve for Na—dissolve 0.254g of AR grade NaCl in distilled water, and make up to 1 liter. This gives 100ppm of Na solution. From this, prepare a series of concentrations of Na (10, 20, 30, 40, 50, 60, 70, 80, 90ppm) by transferring 10, 20, 30, 40, 50, 60, 70, 80, 90mL of 100ppm solution into a 100mL volumetric flask. After inserting the Na filter and regulating the appropriate gas and air pressure, set the flame photometer reading at zero for the blank (distilled water) and at 100 for 100ppm of Na solution. Draw the calibration curve by plotting the flame photometer reading against different concentrations (10, 20, 30, 40, 50, 60, 70, 80, 90, 100ppm of Na).

DOI: 10.1201/9781003430100-25

## 25.4 PROCEDURE

- Put 10g of air-dry soil in a 250mL beaker and add 50mL of neutral normal ammonium acetate solution.
- Stir the contents well and keep it overnight, covering with a watch glass.
- Transfer the soil to the filter paper (Whatman no. 3) and leach it with 25–30mL portions of ammonium acetate solution 6–8 times, collecting the filtrate in a beaker.
- Transfer the filtrate to a 250mL volumetric flask and make up the volume to 250mL.
- Feed the filtrate into the flame photometer for finding the concentration of sodium present in the filtrate and the exchangeable sodium in the soil is calculated accordingly.

## 25.5 OBSERVATION AND CALCULATION

| | | |
|---|---|---|
| Weight of the soil taken | = | 10g |
| Volume of the filtrate prepared | = | 250mL |
| Concentration of sodium in the filtrate as derived from the standard graph for Na | = | Y ppm |
| ∴ Amount of exchangeable sodium in the given soil | = | (Y / 1000) × 250 × (100 / 10) × (1/23) meq / 100 g of soil or C.mol($p^+$) / kg of soil |

## 25.6 THINGS TO LEARN

- What is the purpose of adding neutral normal ammonium acetate?
- What are settings to be changed in the flame photometer, during the estimation for sodium?
- How to convert ppm into mg/lit and meq/100g of soil or C.mol($p^+$)/kg of soil?

# 26 Determination of Exchangeable Calcium and Magnesium in Soil

## 26.1 PRINCIPLE

The method is based on the principle that the soil is first saturated with neutral normal ammonium acetate solution and the excess ammonium ions are removed by washing with alcohol. The displaced cations including calcium and magnesium are collected as the filtrate in a volumetric flask during the estimation of cation exchange capacity of the soil.

The calcium and magnesium present in the filtrate is then estimated by titration with EDTA (versenate). But the ammonium acetate and organic matter must be entirely removed from the soil extracts prior to the titration with EDTA.

Evaporation of an aliquot of the soil extract to dryness followed by treatment with aqua regia (HCl and $HNO_3$ in the ratio of 3:1), and a second evaporation to dryness usually suffices for the removal of ammonium acetate and organic matter. Very dark colored soil extracts may require additional treatment with aqua regia. After the treatment, the residue is dissolved in a quantity of distilled water equal to the original volume of the aliquot taken for treatment.

## 26.2 APPARATUS AND MATERIALS REQUIRED

- Porcelain basin
- Burette
- Conical flask
- Pipette
- Volumetric flask
- Beaker
- Glass rod
- Shaking machine
- Water bath/steam bath

## 26.3 REAGENTS REQUIRED

- 0.02*N* EDTA
- 10% sodium hydroxide

DOI: 10.1201/9781003430100-26

- Ammonium chloride–ammonium hydroxide buffer solution—dissolve 67.5g of ammonium chloride in 400mL distilled water, to this add 570mL of concentrated ammonium hydroxide and dilute to 1 liter
- Murexide indicator
- Eriochrome black-T indicator
- Aqua regia
- 0.1*N* acetic acid
- Red litmus paper

## 26.4 PROCEDURE

### 26.4.1 Preparation of Ammonium Acetate Extract

- Put 10g of air-dry soil in a 250mL beaker and add 50mL of neutral normal ammonium acetate solution.
- Stir the contents well and keep it overnight, covering with a watch glass.
- Transfer the soil to the filter paper (Whatman no. 3) and leach it with 25–30mL portions of ammonium acetate solution 6–8 times, collecting the filtrate in a beaker.
- Transfer the filtrate to a 250mL volumetric flask and make up the volume to 250mL.

### 26.4.2 Pretreatment of Soil Extract

- Transfer the ammonium acetate extract to a 250mL beaker and evaporate to dryness on a hot plate or steam bath.
- Wash down the walls of the beaker with a small quantity of water and again evaporate to dryness.
- Add 1mL of nitric acid and 3mL of hydrochloric acid and again evaporate.
- Dissolve the residue after evaporation in 20mL of 0.1*N* acetic acid.
- Filter through Whatman filter paper (with low ash content) into a 100mL volumetric flask using water to wash the beaker and filter paper and make up the volume to 100mL.

### 26.4.3 Determination of Calcium Alone

- Pipette out 10mL of the aliquot into a porcelain basin.
- Add 10% sodium hydroxide solution drop by drop to neutralize the acidity (red litmus turns blue) and another 5mL excess to maintain the pH at 12.
- Add a pinch of murexide indicator and titrate with 0.01*N* EDTA until the color changes from orange red to lavender or purple. When close to the end point, EDTA should be added at the rate of about a drop every 5–10 seconds, as the color change is not instantaneous.

### 26.4.4 Determination of Calcium + Magnesium

- Pipette out 10mL of the aliquot into a porcelain basin.
- Adjust to pH 10 by adding 15mL of ammonium chloride–ammonium hydroxide buffer (red litmus turns blue).
- Add 3–4 drops of Eriochrome black-T indicator solution and titrate with 0.01*N* EDTA till the color changes from wine red/purple red to sky blue.

## 26.5 OBSERVATION AND CALCULATION

| | | |
|---|---|---|
| Weight of the soil sample taken | = | 10g |
| Volume of the ammonium acetate extract taken | = | 100mL |
| Volume of aliquot (from soil extract/water sample) taken for analysis | = | 10mL |
| ***Calcium*** | | |
| Volume of 0.01*N* EDTA used for calcium alone (titer-value) | = | X mL |
| ∴ Calcium in meq/liter | = | $X \times 0.01 \times (1000/10)$ |
| ∴ Calcium in meq/100g of soil | | $X \times 0.01 \times (1000/10) \times 100 \times (100/10)$ |
| ***Magnesium*** | | |
| Volume of 0.01*N* EDTA used for calcium + magnesium | = | Y mL |
| Volume of 0.01*N* EDTA used for magnesium alone | = | (Y-X) mL |
| ∴ Magnesium in meq/liter | = | $(Y-X) \times 0.01 \times (1000/10)$ |
| ∴ Magnesium in meq/100g of soil | | $(Y-X) \times 0.01 \times (1000/10) \times 100 \times (100/10)$ |

## 26.6 THINGS TO LEARN

- What is complexometric titration? How it helps in the determination of Ca and Mg?
- Which reagent is called versenate? Why it is called so?
- How 1mL of 0.01*N* EDTA = 0.00012g of Mg and 0.0002g of Ca?
- Give the indicators used and end point in the determination of Ca and Mg.

# 27 Determination of Lime Requirement of Soil

## *[Shoemaker, McLean and Pratt (SMP) Method, 1961]*

### 27.1 INTRODUCTION

For proper plant growth, the soil should have a pH between 6.5 and 7.5, although there are certain plants that can grow satisfactorily at low pH (e.g., tea) and at high pH (e.g., sugar beet). In India, acid soils are located mostly in eastern, southern and south central parts, although some soils, at higher elevations, in north India are acidic.

In order to achieve maximum yield and sustained productivity through efficient soil management practices, it is essential to lime an acid soil, as it has considerable influence on soil environment, besides correcting soil acidity.

Several methods have been proposed to determine the lime requirement of acidic soils, to raise the pH to around 6.5. The soil pH value alone, however, is not a good criterion for lime recommendation, because of variations in the exchange acidity of soil, and the nature of crops which may not require as high as this pH to produce maximum yield. Of these various methods proposed, the Shoemaker/SMP method was found satisfactory.

The recent trend of liming is to use it as a fertilizer rather than as amendment so as not to disturb heavily the natural environment while ensuring sustained productivity.

### 27.2 PRINCIPLE

The method involves equilibrating the soil with a pH 7.5 buffer solution, whereby the reserve $H^+$ is brought into solution, which results in the depression of pH of the buffer solution, a note of which is made and interpreted in terms of the lime required to raise the pH to a desired value.

### 27.3 APPARATUS AND MATERIALS REQUIRED

- Beaker
- pH meter

DOI: 10.1201/9781003430100-27

## 27.4 REAGENTS REQUIRED

- Extractant buffer—dissolve 1.8g p-nitrophenol, 3g potassium chromate, 2g calcium acetate, 53.1g calcium chloride dehydrate and 2.5mL triethanolamine in 1 liter of distilled water, and adjust the pH to 7.5 with NaOH.

## 27.5 PROCEDURE

- First of all, determine the pH of the original soil in 1:2.5 soil:water ratio. If the value exceeds 6.0, the method is not applicable to such a soil.
- To maintain 1:2.5 soil:water ratio, take 10g of soil and add 25mL of water and stir the suspension for about half an hour.
- If the pH measured, thereafter, happens to be 6.0 or low, then proceed with the following steps:
- Weigh 5g of soil in a beaker and add 5mL distilled water and 10mL of the buffer. Stir continuously for ten minutes or intermittently for 20 minutes.
- Determine the soil pH with the pH meter.
- Lime requirement is done on the basis of soil-buffer pH ready reckoner table given below:

| pH of soil buffer suspension | Lime required to bring the soil to indicated pH (tons/acre of pure $CaCO_3$) | | |
|---|---|---|---|
| | pH 6.0 | pH 6.4 | pH 6.8 |
| 6.7 | 1.0 | 1.2 | 1.4 |
| 6.6 | 1.4 | 1.7 | 1.9 |
| 6.5 | 1.8 | 2.2 | 2.5 |
| 6.4 | 2.3 | 2.7 | 3.1 |
| 6.3 | 2.7 | 3.2 | 3.7 |
| 6.2 | 3.1 | 3.7 | 4.2 |
| 6.1 | 3.5 | 4.2 | 4.8 |
| 6.0 | 3.9 | 4.7 | 5.4 |
| 5.9 | 4.4 | 5.2 | 6.0 |
| 5.8 | 4.8 | 5.7 | 6.5 |
| 5.7 | 5.2 | 6.2 | 7.0 |
| 5.6 | 5.6 | 6.7 | 7.7 |
| 5.5 | 6.0 | 7.2 | 8.3 |
| 5.4 | 6.5 | 7.7 | 8.9 |
| 5.3 | 6.9 | 8.2 | 9.4 |
| 5.2 | 7.4 | 8.6 | 10.0 |
| 5.1 | 7.8 | 9.1 | 10.6 |
| 5.0 | 8.2 | 9.6 | 11.2 |
| 4.9 | 8.6 | 10.1 | 11.8 |
| 4.8 | 9.1 | 10.6 | 12.4 |

- The values in the table are given in tons of pure $CaCO_3$ per acre, required to bring the soil to the pH indicated, and are required to be converted to their equivalents in the form of agricultural lime to be used.
- For expressing in metric units, i.e., tons per hectare, the figures are to be multiplied by 2.43.

## 27.6 OBSERVATION AND CALCULATION

| | |
|---|---|
| Original soil pH | = |
| Soil pH after equilibration with buffer | = |
| pH of the soil required to be attained after liming | = |
| Amount of lime in terms of pure $CaCO_3$ (tons/acre) | = |
| Amount of lime in terms of pure $CaCO_3$ (tons/ha) | = |

## 27.7 THINGS TO LEARN

- What are the liming materials used for acid soil reclamation?
- What is CCE? Give the CCE for various liming materials.
- Why is gypsum not recommended for the reclamation of acid soils?
- What is a neutralizing index?
- What is the relationship between the neutralizing index and neutralizing value?
- What is the chemical reaction involved in reclaiming the acid soils using lime?

# 28 Determination of Gypsum Requirement of Soil

## *[Schoonover (1952) Method]*

## 28.1 INTRODUCTION

Sodic (alkali) and saline-sodic soils are characterized by the presence of large amounts of sodium, as high as 15% or more, on their exchange complex. As a result of this, the pH of soils increases beyond 8.0, causing nutritional imbalances, elimination of soil organic matter, deterioration of soil physical condition, etc., besides affecting the soil biotic community. Gypsum ($CaSO_4.2H_2O$) is commonly used for the management of such soils, when applied in the right amounts.

## 28.2 PRINCIPLE

A given weight of soil is equilibrated with a known amount of Ca solution, and the amount of Ca left in the solution is determined by EDTA titration. The difference between the amount of Ca added and the Ca left in the solution, gives the amount of Ca exchanged. In practice, gypsum of about 1/3 of the value, obtained by this method, is satisfactory in most cases.

## 28.3 APPARATUS AND MATERIALS REQUIRED

- 50mL burette
- 5mL and 100mL pipettes
- 100mL and 250mL conical flasks
- Filter paper
- Mechanical shaker

## 28.4 REAGENTS REQUIRED

- Saturated $CaSO_4$ solution—shake about 5g of $CaSO_4.2H_2O$ with 1 liter of distilled water for ten minutes, on a mechanical shaker, and filter.
- Ammonium chloride hydroxide buffer—dissolve 67.5g of $NH_4Cl$ in 570 mL of $NH_4OH$ (specific gravity, 0.86), and dilute to 1 liter.
- Eriochrome black-T (EBT) indicator—dissolve 0.5g of EBT and 4.5g of hydroxylamine hydrochloride in 100mL of 95% ethanol.

DOI: 10.1201/9781003430100-28

4. Standard EDTA solution (0.01 *N*)—dissolve 2g of disodium dihydrogen-ethylene-diamine-tetra acetate and 0.05g of $MgCl_2.6H_2O$ in water, and dilute to 1 liter with distilled water. Standardize the solution against the standard $CaCl_2$ solution.

## 28.5 PROCEDURE

- Weigh 5g of the processed soil in a 250mL conical flask, and add 100mL of the saturated $CaSO_4$ solution.
- Shake for five minutes on a mechanical shaker, and filter.
- Pipette out 5mL of the soil extract into a 100mL conical flask, and dilute to about 25mL with distilled water.
- Add 0.5mL of the $NH_4Cl$-$NH_4OH$ buffer and 3–4 drops of the EBT indicator.
- Titrate with the standard EDTA solution until the color changes from wine red to blue.
- Titrate similarly mL of the saturated $CaSO_4$ solution separately.

## 28.6 OBSERVATION AND CALCULATION

| | |
|---|---|
| Weight of the soil taken | = 5g |
| Volume of gypsum solution added | = 100mL |
| Volume of 0.01 *N* EDTA used for gypsum solution | = X mL |
| Volume of 0.01 *N* EDTA used for 5mL of the filtrate | = Y mL |
| Actual volume of 0.01 *N* EDTA used | = (X – Y) mL |
| 1 mL of 0.01 *N* EDTA | = 0.0002g of Ca |
| ∴ (X – Y) mL of 0.01 *N* EDTA | = (X – Y) × 0.0002g of Ca |
| This is present in 5mL of filtrate | |
| ∴ in 100mL of filtrate | = (X – Y) × 0.0002 × (100/5) g of Ca |
| This is present in 5g of soil | |
| ∴ in 2 million kg of soil | = (X – Y) × 0.0002 × (100/5) × [(2 × $10^6$) / 5] g of Ca |
| | = (X – Y) × 1600kg of Ca / ha |
| In terms of gypsum requirement | = 1.6 (X – Y) × (172 / 40) |
| | = 6.88 (X – Y) tons / ha |

## 28.7 THINGS TO LEARN

- How will you reclaim the alkali/sodic soils?
- What is the principle involved in the determination of gypsum requirement?
- What is the chemical reaction involved in reclaiming the sodic soils using gypsum?
- Give the other materials that can be used for the reclamation of sodic soils.

# 29 Enumeration of Microorganisms in Soil

## *Preparation of Serial Dilution*

### 29.1 INTRODUCTION

Quantitative and qualitative estimation of soil microorganisms is necessary to assess the activity of different microbiological ecosystems, operating in the soil which, in turn, helps in understand the mineral nutrition of plants. For this, it is necessary to prepare the serial soil dilutions, from where appropriate dilutions may be taken u for soil microbiological studies.

### 29.2 APPARATUS AND MATERIALS REQUIRED

- Analytical balance
- Sterile pipette
- 250mL flask
- Aluminum can
- Hot air oven
- Desiccator

### 29.3 PROCEDURE

- Weigh 10g of the test soil sample under aseptic conditions in a sterile paper.
- Transfer the sample to 90mL of sterile distilled water.
- Shake thoroughly for uniform mixing, and from this suspension, transfer a 1mL portion, with the help of a sterile pipette, to a 250mL flask, containing 99mL of sterile distilled water.
- Shake the flask thoroughly. This is the 10 times dilution.
- From this, further dilutions can be made by transferring 10mL to 90mL of sterile distilled water aseptically to any desired level, say $10^{-4}$, $10^{-5}$, $10^{-6}$, and so on.
- To determine the water content of the soil, transfer 10g of the soil sample taken in an aluminum can provided with a cover, to an electric hot air oven run at 105°C, and take the weight of the sample at regular intervals.

DOI: 10.1201/9781003430100-29

- When the weight of two intervals attains a constant value, the sample should be transferred to a desiccator for cooling.
- Subsequently take the weight of soil with aluminum can and express the loss, in water, on percentage basis.

## 29.4 OBSERVATION AND CALCULATION

| | | |
|---|---|---|
| Weight of the empty aluminum can | = | X g |
| Weight of the empty aluminum can + moist soil | = | Y g |
| Weight of the empty aluminum can + oven-dry soil | = | Z mL |
| Percentage of water in the soil sample | = | $[(Y - X) / (Z - X)] \times 100$ |

## 29.5 THINGS TO LEARN

- Why do we need to ensure the weight of the soil after two intervals attains a constant value?
- What is serial dilution technique?
- Why do we use sterile distilled water?
- What is the need of maintaining the aseptic condition?

# 30 Composition and Preparation of Microbiological Media

## 30.1 INTRODUCTION

A medium is a mixture of various nutrients, in appropriate amounts, need for the growth of microorganisms. Different types of media are synthetic (the exact chemicals composition is known; for example, Jenson's N-free medium) and non-synthetic or organic medium (exact chemical composition of the some of the nutrients is not known; for example, nutrient agar, soil extract agar, yeast extract mannitol agar and tryptone agar medium).

## 30.2 FUNCTIONS OF DIFFERENT NUTRIENTS

- Yeast extract, beef extract, peptone, tryptone, etc., serve as source of protein, amino acids, vitamins, and growth factors.
- Sugars such as glucose, sucrose, mannitol (sugar alcohol) are used as carbon sources.
- Potassium nitrate, ammonium nitrate, etc., are the source of nitrogen.
- Na, K, Ca, Mg, S, and P are the important constituents of enzymes, nucleic acids, etc. Some of these elements, *viz.*, Na and K also play an important role in maintaining the osmotic balance of the cell.
- Agar-agar serves as the solidifying agent and a medium without agar is called broth.

## 30.3 MATERIALS AND REAGENTS REQUIRED

- Analytical balance
- Beaker
- Sterile pipette
- 250mL flasks/tubes
- Cotton
- Hot air oven
- 0.1*N* HCl
- 0.1*N* NaOH
- Other reagents specified in respective agar medium

DOI: 10.1201/9781003430100-30

## 30.4 COMPOSITION OF GENERALLY USED MEDIA

- **Nutrient agar:** 3g beef extract, 5g peptone, 5g NaCl, 1000mL distilled water, 20g agar-agar, and pH to be maintained is 7.0. This medium is suitable for the growth of most of the bacteria.
- **Soil extract agar:** 100mL soil extract, 0.5g $K_2HPO_4$, 1.8g glucose, 900mL distilled water, 20g agar-agar, and pH to be maintained is 7.0. This medium is suitable for the isolation of most of the soil bacteria. The soil extract shall be prepared by mixing 500g of rich garden soil with 500mL of water, steamed in the autoclave for 30 minutes and subsequently filtering the contents through an ordinary filter paper.
- **Yeast extract mannitol agar:** 10g mannitol, 0.2g $MgSO_4.7H_2O$, 0.1g NaCl, 0.5g $K_2HPO_4$, 1g yeast extract, 1000mL distilled water, 20g agar-agar, and pH to be maintained is 7.0. This medium is suitable for growth of *Rhizobium*.
- **Tryptone agar:** 3g beef extract, 5g tryptone, 1g glucose, 1000mL distilled water, 20g agar-agar, and pH to be maintained is 7.0. This medium is suitable for counting the total number of microorganisms in any food samples.
- **Jenson's N-free agar:** 0.5g $MgSO_4.7H_2O$, 0.5g NaCl, 1g $K_2HPO_4$, 0.005g $FeSO_4$, 0.005g $Na_2MoO_4$, 2g $CaCO_3$, 10g sucrose, 1000mL distilled water, 20g agar-agar, and the pH to be maintained is 7.0. This medium is suitable for the growth of *Azotobacter*.

## 30.5 PROCEDURE

- Weigh the different components of the medium and transfer them to a flask. (Note: Don't add agar-agar at the first instance.)
- Add the required volume of water and dissolve the components.
- Check the pH and adjust it to 7, by using 0.1*N* HCl or 0.1*N* NaOH.
- Add agar-agar and boil the medium to melt the agar.
- Then distribute it in 250mL flask or tubes.
- Plug agar-agar and the flasks/tubes with cotton and sterilize at 120°C for 15 minutes.

# 31 Sterilization Techniques
## *Sterilization of Media and Glassware*

## 31.1 INTRODUCTION

Sterilization is a process by which all microbes, present in any material, are killed. High temperatures, in particular, cause irreversible denaturation of proteins, present in microbes, leading to permanent death.

## 31.2 MATERIALS AND APPARATUS REQUIRED

- Autoclave
- Seitz filters/Millipore membrane filters
- Hot air oven

## 31.3 STERILIZATION TECHNIQUES

### 31.3.1 Sterilization with Steam Under Pressure (Moist Heat)

An autoclave is a double jacketed steam chamber, equipped with devices that permit the chamber to be filled with steam at known pressure and temperature. The advantages of using an autoclave are rapid sterilization, possibility of attaining temperatures above boiling point of water, and no loss of material.

#### 31.3.1.1 Operational Procedure of the Autoclave

- Fill up the chamber with water to the required level.
- Keep the materials to be sterilized on the trays, and close the lid tightly.
- Start heating and allow the steam to pass for five minutes before closing the steam valve.
- Close the outlet valve, allow the pressure to reach 15lb (120°C) and let it stand at that pressure for the desired time of 20 minutes.
- After the lapse of desired time, allow the autoclave to cool.
- When the pressure has come down to zero, slowly open the steam valve, remove the lid of the autoclave and take out the material.
- The autoclave is used for the sterilization of the media, distilled water, normal saline, rubber tubing, etc.

DOI: 10.1201/9781003430100-31

### 31.3.2 Filtration Sterilization

Not all media and solution can be sterilized by heat. Hence, heat labile materials (solutions only) are sterilized and made free from microbes by filtration either through Seitz filters or Millipore membrane filters.

### 31.3.3 Sterilization by Dry Heat: Hot Air Oven

The hot air oven is a dry-air type sterilizer and is used mostly for sterilizing glassware such as tubes, flasks, Petri dishes, etc.

It consists of a chamber with a double wall, insulated with asbestos to prevent the loss of heat. A thermometer is attached to measure the temperature. The hot air oven is operated electrically in the temperature range of 160–180°C for 1–2 hours.

# 32 Determination of Mineralization Rate of Organic Carbon Compounds

## *A Measure of Microbial Activity in Soil*

## 32.1 INTRODUCTION

Carbon is fixed into the organic form by photo-dependent organisms under the influence of light. The fixed C source is not easily available for green plants for continuing the photosynthetic process. The soil microorganisms play a vital role to replenish the C in the atmosphere by decomposing the organic sources in the soil, and thus make it possible for the higher organisms to thrive on the surface of the earth. The C cycle revolves about $CO_2$ fixation and regeneration. Chlorophyll-containing plants utilize the atmospheric $CO_2$ as the sole source of C. The carbonaceous matter, thus synthesized serves to supply the animal world with preformed organic C. The dead plant and animal tissues are acted upon by the soil microorganisms to generate $CO_2$ and thus microbial metabolism plays the dominant role in the above cyclic sequence of fixation and release of $CO_2$ in the atmosphere.

The overall microbial activity of a soil can, therefore, be measured by determining the amount of $CO_2$ evolved from it during a specified period of incubation. The rate of $CO_2$ production in the soil at optimum temperature and water conditions can be measured in the laboratory by absorbing the evolved $CO_2$ in dilute standard Ba $(OH)_2$ solution.

## 32.2 MATERIALS AND REAGENTS REQUIRED

- 250mL conical flasks
- Two-way rubber stopper
- Bent glass tubes
- Aspirator bottle
- 250mL receiver flasks
- $CaCO_3$
- Glucose

DOI: 10.1201/9781003430100-32

- 0.2*N* Ba $(OH)_2$.
- 40% NaOH
- 0.1*N* $H_2SO_4$

## 32.3 PROCEDURE

- Measure 100g of the test soil sample in each of the six 250mL conical flasks and assign the following treatments in duplicate.

  Control
  100mg of $CaCO_3$
  100mg of $CaCO_{3+}$ 1g glucose

- Moisten the soil with tap water and maintain the water level at 80% of the water-holding capacity of the soil.
- Fit the mouths of the conical flasks with two-way rubber stoppers provided with bent glass tubes.
- Connect one end of the bent tube to the bent tube of the 250mL receiver flasks, containing 50mL of 0.2*N* Ba $(OH)_2$ solution. Connect another bent tube of the receiver flask with a rubber tubing to an aspiration bottle, filled with water.
- Connect the other end of the bent tube of the flask, containing soil, to a two-way rubber stopper, fitted to a conical flask, containing 40% NaOH solution, for entrapping the atmospheric $CO_2$, by dipping the air inlet tube into the NaOH solution.
- Arrange all the three flasks in a series and connect them to the aspirator bottle where the air current is controlled at 3–4 bobbles per minute.
- Incubate the flasks for a period of one month or any desired time span. Fill the aspirator bottle as and when necessary.
- Suck the air fast at regular intervals by releasing water from the aspirator bottle rapidly for five minutes and remove the receiver flasks.
- Titrate back the Ba $(OH)_2$ with 0.1*N* $H_2SO_4$. 1mL of 0.2*N* Ba $(OH)_2$ is equivalent to 0.0044g of $CO_2$.
- Besides $CO_2$ evolution, uptake of oxygen into the soil, determination of the decrease in organic matter content in the soil either chemically or by the weight loss method, observation on the degree in specific constituents such as cellulose, lignin, or hemicelluloses, etc., may be taken as the measure of microbial activity in the soil.
- $CO_2$ evolution can be measured by the manometric method where gas exchange in the soil is measured in two respirometer flasks in the presence and absence of alkali. The first flask with alkali detects the $O_2$ uptake, and the second flask detects $O_2$ uptake plus $CO_2$ evolution; the difference between manometers, attached to two flasks, gives the amount of $CO_2$ evolution in a specified time. Dividing the amount of $CO_2$ evolution by the time required for it gives the rate of $CO_2$ production in the soil.

# 33 Estimation of the Nitrifying Power of Soil

## 33.1 INTRODUCTION

The mineralization of organic nitrogenous compound terminates with the production of ammonia, which serves as the starting point of a process, known as "nitrification" which is governed by a specific group of soil microorganisms. Two separate distinct steps are involved in nitrification, where the ammonia is converted into nitrite at first and then nitrite is converted into nitrate. Nitrate production in the soil has considerable agronomic significance. A part of the nitrate produced in the soil, is the major nitrogen source for higher plants and a part of it is subjected to loss from the soil through different pathways such as leaching and denitrification.

## 33.2 APPARATUS AND REAGENTS REQUIRED

- Kjeldahl apparatus for $NH_3$ determination
- Mechanical shaker
- Glass tumbler
- Ammonium sulfate
- Calcium carbonate
- Nessler's reagent—dissolve 50g of KI in a small quantity of cold distilled water. Add to it a saturated solution of $HgCl_2$ until a slight permanent precipitation forms. To this mixture, add 400mL of 50% solution of KOH, after allowing the suspended particles to settle. Keep the reagent in a well-stoppered bottle and store it in a dark place.
- Trommsdorf's reagent for $NO_2$—add slowly, with constant stirring, to a boiling solution of 20g $ZnCl_2$, in 100mL distilled water, a mixture of 4g of starch in water. Continue heating until the starch is dissolved as much as possible and the solution is clear. Dilute the mixture with water and add to it 2g of zinc iodide. Make the volume to 1 liter and filter. Store the reagent in a well-stoppered bottler and keep it in a dark place. Solution of diluted sulfuric acid (1:3) is the complementary test solution. Keep this solution separately.
- Diphenylamine reagent—dissolve 0.7g of diphenylamine in a mixture of 60mL of concentrated sulfuric acid and 28.8mL of distilled water. Cool this mixture

DOI: 10.1201/9781003430100-33

and add slowly 11.3mL of concentrated hydrochloric acid. Let the mixture stand overnight and then it can be used.

- Devarda's alloy and MgO
- 2*N* KCl
- Cellulosic materials
- 0.05*N* NaOH solution
- 0.05*N* $H_2SO_4$
- Methyl red indicator

## 33.3 PROCEDURE

- Weigh 100g of the test soil in each of four glass tumblers and allot the following treatments

  Control
  0.2g of ammonium sulfate
  0.2g of ammonium sulfate + 2.5g cellulosic materials
  0.2g of ammonium sulfate + 2.5g cellulosic materials + 0.5g $CaCO_3$

- Mix the ingredients of different treatments in the soil and then maintain the water content at 50% of the field capacity moisture.
- Determine the pH of each sample of the soil in the tumbler and weigh each of them. Cover all the tumblers with sterile hard board and incubate them at 30±1°C.
- At regular intervals, weigh the tumbler and compensate the loss of water. The volume of water to be added will be indicated by the difference in weights of the tumbler with the original weight. At the end of the second and third weeks, make the qualitative test for the following:

  - $NH_3$—by the use of Nessler's reagent
  - $NO_2$—by the use of Trommsdorf's reagent
  - $NH_3$—by the use of diphenylamine reagent

  At the end of one month, make a quantitative estimate of the following parameters in the incubated soil in the tumbler.

  - pH, $NH_4^+$ by KCl extraction procedure
  - $NO_2^-$ + $NO_3^-$ contents by using Devarda's alloy, after distillation of the incubated soil with MgO

- Put 10–25g of the incubated soil in a 250mL conical flask, add to it 100mL of 2*N* KCl solution and shake vigorously in a mechanical shaker for an hour.
- Allow the soil particles to settle down and filter thorough a Whatman no. 1 filter paper.

- Leach the contents in the filter paper with small amount of 2*N* KCl solution until the filtrate gives no coloration with Nessler's reagent. Normally, five to seven washings are required to make the soil free from $NH_4^+$.
- Transfer the combined extract to a 1000mL Kjeldahl distillation flask, and add a little paraffin to prevent frothing. Add sufficient amount of MgO to make the solution alkaline.
- Distil the contents of the flask and absorb the ammonia released from the extract in 25mL of 0.05*N* $H_2SO_4$ solution.
- Disconnect the flask and allow it cool down for about half an hour.

## 33.4 OBSERVATIONS

- Test the presence of $NH_3$ in the distillate and, when it is free from $NH_3$, remove the receiver flask and titrate with 0.05*N* NaOH solution, using methyl red as the indicator.
- Subtract the titer value from the original volume of 0.05*N* $H_2SO_4$ solution, taken in the receiver flask.
- Distil an equal amount of KCl solution, used from the extraction of the soil with MgO, and absorb the distillate in 25mL of 0.05*N* $H_2SO_4$ solution. Titrate it with 0.05*N* NaOH solution, using methyl red as the indicator. Take the reading of this flask as "control."
- Subtract the titer value from the original volume of 0.05*N* $H_2SO_4$ solution to obtain the volume of the acid used for the neutralization of $NH_3$ released during distillation.
- Add to the leftover filtrate in the distillation flask 0.2g Devarda's alloy and mix thoroughly. Allow the content of the flask to stand for about half an hour and then distil the content for about half an hour. Devarda's alloy reduces the $NO_2^-$ and $NO_3^-$ to $NH_3$, which can be absorbed in a known volume of 0.05*N* $H_2SO_4$. Back-titrate the distillate with 0.05*N* NaOH solution, using methyl red as indicator. Deduct the titer value from the original volume of $H_2SO_4$ solution taken in the receiver flask. This is taken as the amount of acid neutralized by $NH_3$ evolved during distillation. Similarly, distil the control flask, treated with an equal amount of Devarda's alloy and deduct the value of the final value of $NO_2^- + NO_3^-$ obtained from the soil extract.

# 34 Isolation and Quantitative Estimation of *Azotobacter* in Soils

## 34.1 INTRODUCTION

Biological nitrogen fixation is brought out by free living bacteria, blue-green algae, and through symbiotic association between microorganisms and higher plants. The biological nitrogen fixation is the natural process through which nitrogen, removed by various pathways, is replenished for the growth of plants. Although nitrogenous fertilizers are being increasingly used for crop production, biological nitrogen fixation through various microorganisms needs reevaluation, considering the fact that extensive use of fertilizers causes soil and water pollution.

Biological nitrogen fixation in the soil can be quantified by measuring the increase in the nitrogen content of an enrichments culture or of the natural soil by the conventional nitrogen estimation method in the laboratory. However, the estimation of *Nitrosomonas* activity of a medium, inoculated with some nitrogen fixing microorganisms, is the best and accurate method.

The microorganisms capable of fixing atmospheric nitrogen non-symbiotically are *Azotobacter*, *Azomonas*, *Beijerinckia*, *Derxia*, *Methylomonas*, *Mycobacterium*, *Spirillum*, *Bacillus*, *Enterobacter*, *Klebsiella*, *Clostridium*, *Desulfovibrio*, etc.

Among all the bacteria mentioned above, extensive work has been done on *Azotobacter*. *Azotobacter* fixes about 1mg of N per mL of the medium with an efficiency of 5 to 20mg N per gram of sugar oxidized.

## 34.2 ISOLATION OF *AZOTOBACTER*

### 34.2.1 Materials and Reagents Required

- Incubator
- Analytical balance
- Hot air oven
- Sterile Petri plates
- Base medium 77 (0.5 g $K_2HPO_4$, 0.2g MgSO4.7$H_2O$, 0.2g NaCl, trace of $MnSO_4.4H_2O$, trace of $FeCl_3.6H_2O$, and 1000mL distilled water)
- 1.5% agar
- 1% mannitol

DOI: 10.1201/9781003430100-34

### 34.2.2 Procedure

- Prepare a soil dilution by adding 5g of the test soil to 25mL sterile water blank. Shake vigorously and allow to stand for a few minutes.
- Inoculate 1mL of the soil solution to 25mL base medium 77 (0.5 g $K_2HPO_4$, 0.2g MgSO4.7$H_2O$, 0.2g NaCl, trace of $MnSO_4.4H_2O$, trace of $FeCl_3.6H_2O$, and 1000mL distilled water), containing 1mL of mannitol solution.
- Incubate the flask at room temperature, if the ambient temperature is optimum, otherwise incubate the flask in the incubator, maintained at 25–30°C.
- Examine the inoculated flask at regular intervals, taking a portion of the film appearing on the surface under oil immersion objective. When *Azotobacter* cell growth is observed in the flask, prepare loop dilutions from the enrichment culture in a series of 9mL water blank.
- Inoculate a few sterile Petri plates with 1mL of selected loop dilutions, and pour sterile melted base medium 77, containing 1.5% agar and 1% mannitol.
- Incubate the Petri plates at 25–30°C in an incubator and examine at regular intervals for typical *Azotobacter* colonies.
- Transfer a portion of the growth from typical colonies to slants prepared with base medium 77, containing 1.5% agar and 1% mannitol.
- Purify the culture by frequent transfer and the loop dilution technique. Preserve the culture for subsequent studies.

## 34.3 QUANTITATIVE ESTIMATION OF *AZOTOBACTER* IN SOILS

### 34.3.1 Procedure

- Collect a representative soil sample from the field, where a quantitative estimation of *Azotobacter* count is to be obtained.
- Weigh approximately 10g of fresh soil and suspend it in 90mL of sterile distilled water in a bottle.
- Make serial soil dilutions from this and preserve the selected dilution for quantitative studies.
- Transfer an equal amount of the soil sample to an electric hot air oven, maintained at 105°C for determining the water content of the soil.
- Inoculate 1mL of the selected soil dilution to each of three sterile Petri plates, and pour melted base medium 77 with 1.5% agar and 1% mannitol at 47°C.
- Mix the soil dilution thoroughly with the medium by the usual procedure. Allow the Petri plates to cool down for solidification of the agar medium, and then transfer them to an incubator maintained at 25–30°C.
- Examine the Petri plates after 72 hours and take counts of *Azotobacter* colonies in the Petri plates. Transfer a portion of the growth from typical colonies to slants, prepared with the same medium, for subsequent microscopic and culture studies.

# 35 Isolation and Quantitative Estimation of *Rhizobia* in Soils

## 35.1 INTRODUCTION

Only a fraction of the nitrogen needed for agricultural production comes from synthetic sources, the remaining portion comes from the soil reserve and biological nitrogen fixation of atmospheric $N_2$. Some amount of $N_2$ is fixed in the soil during lightning. Biological nitrogen fixation is accomplished by both free-living microorganisms and symbiotic processes.

Some plant species, including the lower form of life, establish symbiotic associations with certain microorganisms for the acquisition of $N_2$ from the atmosphere. Leguminous plants are a classic example of such a process where the bacteria of the genus *Rhizobium* act as the microsymbiont. The microorganisms initiate the formation of root nodules, which is the site for the symbiotic process. The microsymbiont fixes the atmospheric $N_2$ at this site, and thus fulfils the requirement of the macrosymbiont, which in turn supplies carbohydrates to the bacteria for their growth and $N_2$ fixation.

Legumes belong to the family fabaceae, which are divided into three subfamilies. In the two-family, "Papilionoideae" genus, *Trifolium*, *Melilotus*, *Medicago*, *Phaseolus*, *Crotalaria*, *Vicia*, *Vigna*, *Pisum*, and *Lathyrus* are important plants in agriculture.

## 35.2 ISOLATION OF *RHIZOBIA* FROM ROOT NODULES

### 35.2.1 Materials and Reagents Required

- Camel hair brush
- Sterile forceps
- Petri dish/Petri plates
- Sterile glass rod
- Test tube
- Incubator
- Nodule of the test plant
- 1:1000 parts of $HgCl_2$ solution
- 75% ethyl alcohol

DOI: 10.1201/9781003430100-35

- Yeast extract mannitol medium (10g mannitol, 0.5g $K_2HPO_4$, 0.2g $MgSO_4.7H_2O$, 0.1g NaCl, 3g $CaCO_3$, 100mL yeast extract (10%) solution, 15g agar, 900mL distilled water)

### 35.2.2 Procedure

- Carefully remove a nodule from the root of the test plant, keeping a portion of the root intact to the nodule.
- Wash the nodule under running water with a camel hair brush to dislodge any soil particles adhering to it.
- Place the nodule in a Petri dish, containing 1:1000 parts of $HgCl_2$ solution for 3–6 minutes, so that the nodule is completely submerged in $HgCl_2$ solution.
- Remove the nodule from the Petri dish, with the help of a sterile forceps and transfer it to a Petri dish containing 75% ethyl alcohol.
- Agitate the nodule in the Petri dish for one minute. Transfer the nodule to a Petri dish, containing sterile water and agitate to wash it completely.
- Finally, transfer the nodule to a test tube, containing a few milliliters of sterile distilled water. Crush the nodule in the test tube with the help of a sterile glass rod.
- From this initial suspension of nodule bacteria, make serial dilutions by transferring one loopful to 10mL of sterile water, taken in a test tube. Continue the process at least for three dilutions.
- From the second and third dilutions, transfer 1mL to each of the three sterile Petri plates, maintained for each dilution.
- Pour the melted yeast extract mannitol (10g mannitol, 0.5g $K_2HPO_4$, 0.2g $MgSO_4.7H_2O$, 0.1g NaCl, 3g $CaCO_3$, 100mL yeast extract (10%) solution, 15g agar, 900mL distilled water) at 45°C and thoroughly fix the inoculums with the media by shaking several times crosswise, clockwise, and anti-clockwise.
- Allow the agar medium in the Petri dish to solidify and then transfer them to an incubator, maintained at 25–30°C and incubate for 72 hours.

### 35.2.3 Observations

- At the end of the incubation period, examine the Petri plates for the growth of *Rhizobium* bacteria.
- Select two isolated *Rhizobium* colonies and transfer one loopful to slants, prepared with the same medium.
- Examine the bacteria under a microscope, and perform all cultural and staining reaction tests.

# 36 Isolation and Purification of Ectomycorrhizal Fungi

## 36.1 INTRODUCTION

An ectomycorrhiza is a root/fungus association, where the fungus is a mantle on the surface of the root, and penetrates the cortex intercellularly to produce a network called a "Hartig net." Intercellular infection occurs in the cortical cells and the composite organ is termed an "ectendomycorrhiza." Ectomycorrhizae display a wide range of shape, color, and size, depending upon the host and the fungal species. Ectendomycorrhizae are somewhat less diverse in morphology.

Ectomycorrhizae and ectendomycorrhizae develop between an active fungus and a developing feeder root of a woody plant, and environmental factors affect the quantitative nature of association.

The recognition of the modification induced by mycorrhiza may be simple. If it is compared with a non-mycorrhizal root system under favorable conditions or under normal environmental conditions.

Ectomycorrhizae appear from one to three months after germination. In most coniferous nurseries, mycorrhizae develop spontaneously. Initial delay in mycorrhizal formation may lead to the escape of some roots from infection in the first year. During the second year, mycorrhizal development proceeds rapidly and practically all short roots become mycorrhizal. Studies on mycorrhizal fungi, from coniferous nursery seedlings, have shown that the ectomycorrhizae are formed by relatively few fungal species in contrast to the delivery found under a natural forest ecosystem. In many forest nursery seedlings of *Pinus* and *Larix*, mycorrhizal association exists. This is characterized by coarse intercellular hyphae, which extends to the cortical cells of long roots. This ectendomycorrhizal association occurs on other members of the *Pinaceae* family.

Cultural studies of ectomycorrhizae provide knowledge on basic biology, processes of fungal symbiont, growth response at different pH, temperature and moisture regimes, mineral and carbohydrate sources, and production of enzymes and hormones. Isolation and comparative study of diverse groups of ectomycorrhizal fungi and isolate, within the species, provide a basis for selecting specific isolates for artificial inoculation of nursery stock.

DOI: 10.1201/9781003430100-36

## 36.2 ISOLATION OF ECTOMYCORRHIZAL FUNGI

Ectomycorrhizal fungi can be isolated from sporocarp tissue, surface sterilized ectomycorrhizae, sclerotia, rhizomorphs, and, in some cases, from sexual spores. Sporocarp is generally preferred for the isolation of ectomycorrhizae, as species can be identified easily, and little pretreatment of fungal material is required. Many species within the following genera, can be isolated from sporocarp—*Alpova, Amanita, Astraeus, Boletus, Cortinarius, Fuscoboletinus, Hebeloma, Hymenogaster, Hysterangium, Laccaria, Lactarius, Leccium, Melanogaster, Paxillus, Pisolithus, Rhizopogon, Scleroderma, Suillus*, and *Tricholoma*.

## 36.3 COLLECTION OF SPOROCARP

Young sporocarps are preferred for direct isolation, although fully matured specimens should be collected for reliable identification of the species. Plant sporocarps into waxed paper bags or wrapped paper to prevent drying (avoid plastic bags). For mushroom fungi, cut the stem from a mature specimen in good condition, place the cap over a piece of white card, cover with waxed paper, and flatten the cap over the card in the collection container. Record macroscopic sporocarp characters, and note potential hosts in the vicinity at the time of collection of spores. Isolation should be done immediately after collection. Properly stored sporocarps of some species may be used for successful isolation of ectomycorrhizal fungi even after several days. During collection, avoid freezing, heating, or drying of sporocarp. If isolation is delayed, the collected sporocarp samples should be stores under refrigerated conditions.

## 36.4 COLLECTION OF ECTOMYCORRHIZAE

Select roots with little bit of adhering debris when collecting the samples. Collect enough materials for the isolation exercise for each ectomycorrhizal type. Isolate immediately after collection. Since fine roots dehydrate quickly, it is advisable to place the root specimen together with moist soil, humus, or moss in a tightly closed can or plastic bag. If the study is delayed, preserve the sample in refrigerated conditions. This procedure should be followed for rhizomorphs also.

Young sporocarps are preferred for direct isolation, although fully matured specimens should be collected for reliable identification of the species. Plant sporocarps into waxed paper bags or wrapped paper

## 36.5 ISOLATION FROM SCLEROTIA

This method is suitable for *Genococcum geophilum*. Sclerotia are found abundantly in some ectomycorrhizal hosts, and they can be isolated from soil samples at any time of the year. To collect sclerotia, take away the upper humus and organic-mineral soil interfaces to a depth of 10–15cm. At least five subsamples should be collected with the help of a trowel, around the immediate vicinity of the host plant, to yield 1–2 liters of soil by volume. Take the samples to the laboratory and store them in a tightly closed container under refrigerated conditions, until sclerotia are isolated. Although the samples can be stored for many weeks without loss of viability of sclerotia, it is advisable to isolate them as soon as possible, after collection of the soil samples from the field.

## 36.6 PREPARATION FOR ISOLATION

Isolation should be done in a clean still-air chamber to avoid contamination. The necessary tools and materials include fine tipped, heat sterilized scalpels, transfer needles, forceps, alcohol, gas lamps or sprit lamps, and general disinfectants, such as 5% sodium hypo chloride or 95% ethanol. Prepare sufficient slants and Petri plates with nutrient agar or potato dextrose agar. Antibiotics should be used to reduce contamination, especially for isolation from roots.

## 36.7 ISOLATION FROM SPOROCARP TISSUE

Select young sporocarp, free from rot and insect damage. Brush off adhering debris from the stipe base. If it is difficult to clean, cut it off. For mushroom, cut a shallow (1–2mm) slit cross the middle of the cap surface. In the case of tuber-like hypogenous sporocarp, the initial shallow cut should go around approximately one-half of the fruit body. Now, gently, pull apart the sporocarp along the initial shallow cut, using fingertip pressure. Cut and loosen small pieces of tissue with the help of a sterile scalpel. Transfer the tissue with the scalpel or a sterile transfer needle directly onto the nutrient agar in test tubes or in Petri plates. Isolate from two or three locations on the sporocarps—cap and stipe. For most hypogenous fungi such as *Rhizopogon sp.*, the center of the gleba is best for isolation, whereas, for *Scleroderma sp.*, the interior of the leathery peridial tissue is good. Six to ten tissue transfers are sufficient for isolation of one species.

After 3–4 days, observe the test tubes or Petri plates under a stereoscopic microscope for fungus growth. Fungi that grow well in culture medium produce visible mycelia growth 4–7 days after isolation. Other fungi may take 2–6 weeks to show visible growth. After establishment of the ectomycorrhizal fungi, on the agar substrate mycelium, from the new colony edge, they should be aseptically transferred onto the fresh nutrient agar, and mark it as "stock culture." Bacteria and several fungi are the most common contaminants and can be easily detected by the experienced worker. Bacterial contaminants can be detected by glistening slime, and fungi by characteristic conidia in the growth medium.

## 36.8 ISOLATION FROM ECTOMYCORRHIZAE

After collection, ectomycorrhizae are stored and the cleanest ones are selected; the following treatments are given:

- Rinse the roots under tap water to remove loosely adhering debris.
- Place the rootlets into a perforated plastic vial, and shake vigorously for three minutes in a mild detergent solution.

- Rinse off detergents with tap water and soak for four minutes in 100ppm mercuric chloride or 5–20 seconds in 30% hydrogen peroxide.
- Rinse immediately in 2 liters of sterile distilled water and aseptically transfer individual ectomycorrhiza onto the nutrient agar in test tubes or Petri plates.
- If the results are not satisfactory, vary the sterilization time.
- The test tubes or the Petri plates, inoculated with surface sterilized ectomycorrhizae, should be incubated. Growth occurs 2–4 weeks after inoculation.

## 36.9 ISOLATION FROM SCLEROTIA

Place small soil samples in an evaporating dish and wet to make a slurry with free water. Swirl the dish so that the low-density soil fraction containing sclerotia floats to the surface. Then pour off into another dish. A large volume of soil can be wet and sieved through a 0.344mm screen. Place final fractions in water under a stereoscopic microscope, and remove black sclerotia with the help of forceps. Then treat sclerotia similar to ectomycorrhizae for isolation. Select large, clean sclerotia free from cracks. Remove the adhering debris, and soak for 10–20 minutes in 30% hydrogen peroxide. Rinse sclerotia in sterile water and aseptically transfer onto the nutrient agar and incubate. Isolate from 30–40 sclerotia, per soil sample, to ensure isolation of the organism. Near-white gray hyphal tips, usually appear two weeks after inoculation. Established colonies are easily recognized as coarse hyphae, which become jet black behind the tip.

## 36.10 ISOLATION FROM SCLEROTIA

For future study, stock cultures should be maintained separately from working cultures. Stock cultures are most commonly stores on the nutrient agar slant, in test tubes, under refrigerated conditions (3–5°C.)

# 37 Collection and Preparation of Plant Samples for Laboratory Analysis

## 37.1 INTRODUCTION

Accumulation of nutrient elements in plant tissues indicates the accessibility of the concerned elements from the soil to the plant. Although different plant species and even different varieties of the same species may vary in their nutrient requirements, composition of a part or of the whole plant may well be adopted as a guideline to support the soil-test results for factual assessment of soil's nutrient supplying power. Also, the concentration of certain heavy metals in plants might be a result of their solubility in soil and may lead to polluting effects.

To a certain extent, there exists a good relationship between the concentration of an element in the plant and the total biomass of the plant. A correct balance of nutrients in the plant tissues is closely associated with the maximum yield, except in the cases of luxury consumption of nutrients like potassium.

Plant analysis as an aid to soil testing has its limitations too. These, among other things, include the risk of sampling error as well as the chances of wrong interpretation of the test results for making recommendations. Significant errors in the plant analysis may arise due to wrong sampling alone, especially when the plant analysis aims at diagnosis of nutrient deficiency in standing crops. Selecting the right plant part, stage of growth, and time of sampling are very crucial in plant analysis.

## 37.2 PLANT SAMPLING

For a meaningful plant analysis, utmost care should be exercised in plant sampling. Whole plant analysis is done for working out the total nutrient uptake, which is usually carried out on the aboveground (shoot) material.

Analysis of roots may either be taken up separately or individual plant parts like leaf, petiole, etc., are sampled for the purpose of diagnosing nutrient deficiency in perennial fruit trees. Thus, depending on the purpose of analysis, plant sampling needs to be planned. The table below shows the specific tissues to be sampled from different plant species.

DOI: 10.1201/9781003430100-37

| Crop | Part to be sampled with stage/age |
|---|---|
| ***Grain, pulses, oilseeds, fiber and commercial crops*** | |
| Wheat | Flag-leaf, before head emergence |
| Rice | Third leaf from apex, at tillering |
| Maize | Ear-leaf, before tasseling |
| Barley | Flag-leaf at head emergence |
| Oat | Flag-leaf, before inflorescence emergence |
| Pulses | Recently matured leaf, at bloom initiation |
| Groundnut | Recently matured leaflets, at maximum tillering |
| Sunflower | Youngest mature leaf blade, at initiation of flowering |
| Mustard | Recently matured leaf, at bloom initiation |
| Soybean | Third leaf from top, after two months of planting |
| Cotton | Petiole, fourth leaf from the apex, at initiation of flowering |
| Jute | Recently matured leaf, at 60 days age |
| Sugarcane | Third leaf from top, after 3–5 months of planting |
| Sugar beet | Petiole of youngest mature leaf, at 50–80 days age |
| Tea | Third leaf from tip of young shoots |
| Coffee | Third or fourth pair of leaf apex of lateral shoots, at bloom |
| ***Vegetables*** | |
| Potato | Most recent, fully developed leaf (half-grown) |
| Tomato | Leaves adjacent to inflorescence (mid-bloom) |
| Onion | Top non-white portion (1/3 to 1/2 grown) |
| Brinjal | Blade of most recent, fully developed leaves |
| Beans | Uppermost, fully developed leaves |
| Cauliflower | Most recent, fully matured leaf, at heading |
| Cabbage | Wrapper leaf at 2–3 months age |
| Pea | Leaflets from most recent, fully developed leaves, at first bloom |
| Carrot | Most recent, fully matured leaf, at mid-growth |
| Radish | Most recent, fully developed leaf |
| Turnip and sugar beet | Most recent, fully developed leaf, at 30–50 days age |
| Spinach | Fifth leaf from tip (omit unfurled) at the stage of bud starting to small fruits |
| Cucumber | Most recent, fully developed leaf |
| ***Ornamental plants*** | |
| Bougainvillea, jasmine, croton, fern, ficus, geranium, gerbera, gladiolus, lily, orchid, hibiscus and rose | Most recent, fully developed leaves at the stage of flower bud of pea size |
| ***Fruit crops*** | |
| Almond | Third leaf from top, at the beginning of bloom |
| Apple, pear | Leaves from middle of terminal shoot growth, 8–12 weeks after full bloom, 2–4 weeks after formation of terminal buds in bearing trees |
| Blackberry | Latest matured leaf from non-tipped canes, 4–6 weeks after peak bloom |
| Cheery | Fully expanded leaves, mid shoot current growth in July–August |

| Crop | Part to be sampled with stage/age |
|---|---|
| Peach | Mid-shoot leaves, fruiting or non-fruiting spurs, mid-summer leaves, fruiting |
| Strawberry | Youngest, fully expanded matured leaf without petiole, at peak or harvest period |
| Plum | Leaves form middle of current season's extension growth, in January–February |
| Banana | Petiole of third open leaf from apex, four months after planting |
| Cashew | Fourth leaf from tip of matured branches, at beginning of flowering |
| Grapes | Fifth petiole from base at bud differentiation for yield, and petiole opposite to bloom for quality |
| Citrus fruit | 3–5-month-old leaves from new flush, first leaf of the shoot, in June |
| Guava | Third pair of recently matured leaves, at bloom (August or December) |
| Mango | Leaf with petiole (4–7 months old) from middle of shoot |
| Papaya | Sixth petiole from apex, six months after planting |
| Pineapple | Middle third portion of white basal portion of fourth leaf from apex, at 4–6 months age |
| Pomegranate | Eighth leaf from apex at bud differentiation, in April and August |
| Falsa | Fourth leaf from apex, one month after pruning |
| Ber | Sixth leaf from apex secondary or tertiary shoot, two months after pruning |

## 37.3 PROCEDURE FOR PLANT SAMPLING

- For analysis of seasonal crop plants, pick up few representative plants at random from each plot, remove the shoot (aerial parts) with the help of a sharp stainless-steel knife, blade, or scissors for whole-shoot analysis or the desired part for analysis of specific plant parts.
- If roots are to be included, uproot the whole plant carefully from wet soil, retaining even the fine/active roots.
- Gently dip the plant roots in water several times to remove adhering soil as far as possible.
- Wash the plant parts with water several times.
- Wash the samples thoroughly with about 0.2% detergent solution to remove the waxy/greasy coating on the leaf surface, which is often present.
- Wash with 0.1 *N* HCl followed by a thorough washing with plenty of water. Rinse with distilled water.
- Rinse with double distilled water, especially if micronutrient analysis is to be carried out.
- Soak dry with good quality tissue paper.
- Air-dry the samples on a perfectly clean surface at room temperature for at least 2–3 days in a dust-free atmosphere away from any kind of contaminants.
- Place the samples in an electric oven and dry at 70°c for 48 hours.

- Grind the samples in an electric stainless-steel mill using a 0.5mm sieve.
- Clean the cup and blades of the grinding mill before each sample.
- Place the samples back in the oven and dry again for few hours more for constant weight.
- Store in well-stoppered plastic or glass bottles or paper bags for analysis.

## 37.4 THINGS TO LEARN

- What is index tissue?
- List out the plant parts to be sampled for the commonly grown horticultural crops in your region.
- List out the various plant parts, other than leaves, to be sampled.
- How do the processed plant samples need to be stored?

# 38 Determination of Total Nitrogen in Plants *(Micro Kjeldahl Method)*

## 38.1 INTRODUCTION

Nitrogen (N) is one of the major constituents required for the nutrition of plants. It plays an important role in synthesis of proteins, which are responsible for various metabolic activities in plants. Nitrogen is absorbed by plants in large quantities for their growth and development, the absorption being in the grand growth of plants. Nitrogen content of plants varies from 0.2% to 6% of dried material.

## 38.2 IMPORTANCE

A knowledge of N content, in plant tissues, is important to diagnose its deficiency or toxicity in plants, based on its critical levels, and to make appropriate ameliorative measures, in terms of nutrient management, for increased crop production.

## 38.3 PRINCIPLE

A known weight of powdered sample is treated with diacid mixture so as to oxidize the organic matter and bring the mineral elements into a solution wherein the organic form of nitrogen is converted into ammonium sulfate. The digested material is distilled with excess alkali and liberated ammonia is absorbed in boric acid. The amount of N present in the sample is calculated by titrating the ammonia liberated with standard sulfuric acid.

## 38.4 APPARATUS AND MATERIALS REQUIRED

- 100mL conical flask
- 250mL volumetric flask
- Funnel
- 10mL measuring cylinder
- 25mL pipette
- Burette
- 250mL beaker/ice tumbler
- Distillation set

DOI: 10.1201/9781003430100-38

- Sand bath
- Balance

## 38.5 REAGENTS REQUIRED

- Diacid mixture (sulfuric acid and perchloric acid in the ratio of 5:2)
- 2% boric acid
- *N*/10 $H_2SO_4$
- Bromocresol green–methyl red double indicator
- 40% NaOH
- Red litmus paper

## 38.6 PROCEDURE

- Transfer 1g of plant sample into a 100 mL conical flask.
- Add about 10–15mL of diacid mixture (sulfuric acid and perchloric acid in the ratio of 5:2) and cover the mouth of the flask with a funnel.
- Digest the contents of the flask, over a sand bath till a clear solution is obtained.
- Transfer the contents of the flask into a 250mL volumetric flask using minimum quantity of distilled water and make up the volume to 250mL.
- Pipette out 10mL of the diacid extract into a distillation flask.
- Pipette out 25mL of 2% boric acid into an ice tumbler/250mL beaker and add two drops of double indicator (bromocresol green and methyl red) and keep it at the delivery end of the distillation set.
- Add 10mL of the 40% NaOH to the distillation flask with a moistened red litmus paper and start distillation.
- Absence of blue color of litmus paper indicates that all ammonia has been distilled.
- Then wash the delivery tube with distilled water and collect the washings in the ice tumbler/beaker.
- Titrate the contents with *N*/10 $H_2SO_4$ and the end point is the change of color from blue to green.

## 38.7 OBSERVATION AND CALCULATION

| | | |
|---|---|---|
| Weight of the sample taken | = | 1.0g |
| Volume of the diacid extract prepared | = | 250mL |
| Volume of diacid extract pipetted out for analysis | = | 10mL |
| Volume of *N*/10 $H_2SO_4$ consumed | = | X mL |
| 1mL of *N*/10 $H_2SO_4$ | = | 0.0014g of N |
| ∴ X mL of *N*/10 $H_2SO_4$ | = | 0.0014 × X g of N |
| This is present in 10mL of the diacid extract | | |
| ∴ in 250mL of the diacid extract | = | 0.0014 × X × (250 / 10) |
| This is present in 1g of the plant sample | | |
| ∴ in 100g/percentage of nitrogen (% N) in the given plant sample on moisture free basis | = | 0.0014 × X × (250 / 10) × (100 / 1) × (100/100-M)<br>[M = moisture content of the sample] |

## 38.8 INTERPRETATION

| N content in % of dried material (cereals) | Comments |
|---|---|
| < 1.0 | Deficiencies with symptoms |
| 2.0–4.0 | Optimum supply zone |
| > 5.0 | Toxicity |

## 38.9 THINGS TO LEARN

- Why is diacid extraction adopted for total N?
- What is the end point of the titration in the estimation of N?
- Give the critical level of N in plants.
- Describe the deficiency and toxicity symptoms of N in plants. How will you correct the deficiency and toxicity symptoms?

# 39 Determination of Total Phosphorus in Plants

## *(Colorimetric/ Vanadomolybdate Yellow Color Method)*

### 39.1 INTRODUCTION

A plant analysis is used as a diagnostic tool for identifying the plant nutrient status at any point of time. It is based on the concept that the concentration of an essential element in a plant or part of the plant indicates the soil's ability to supply that nutrient.

Thus, nutrient concentrations in plants are assumed to be directly related to the quantity of that nutrient in the soil that is available to the plant. A second assumption is that, up to a certain critical point, the content of a nutrient in the plant is directly related to yield.

Plant analysis, usually refers to the quantitative analysis for the total amount of essential elements in a plant tissue. A plant removes a large quantity of nutrients from the soils and applied fertilizers, and this removal is reflected in growth and yield of a crop. The average range of variation of phosphorus (P) in the plant is 0.05% to 2%.

### 39.2 IMPORTANCE

Plant analysis is absolutely necessary in order to examine the nutrition attentively with the view to establishing deficiencies, abundance, or excess and to applying agro-chemical measures, so that the critical level should approach the normal condition.

### 39.3 PRINCIPLE

A known weight of a sample is digested with triacid mixture to bring the mineral constituents into a solution. The amount of phosphorus present in the sample is estimated based on the intensity of yellow color developed with vanadomolybdate in the nitric acid medium.

When vanadomolybdate and phosphorus radical react in nitric acid medium, a heteropoly compound is formed that is yellow in color. The intensity of yellow color is proportional to the amount of phosphorus in the sample. By reading the intensity of yellow color produced in a spectrophotometer at 400–490nm, the amount of phosphorus in the sample is determined.

DOI: 10.1201/9781003430100-39

## 39.4 APPARATUS AND MATERIALS REQUIRED

- Volumetric flask of various capacities
- Spectrophotometer
- 5mL pipette
- Funnel
- Balance

## 39.5 REAGENTS REQUIRED

- Barton's reagent—dissolve 25g of ammonium molybdate in 400mL water and mix this with a solution containing 1.25g of ammonium meta vanadate in 300mL concentrated nitric acid and make up the volume to 1 liter.
- $KH_2PO_4$ (AR grade)
- Triacid extract
  - Weigh 1g of the sample into a 250mL conical flask and add 12–15mL of triacid mixture (nitric acid:sulfuric acid:perchloric acid at 9:2:1 ratio) and cover the mouth of the flask with funnel.
  - Digest the contents over a sand bath at 180–200°C until a clear solution is obtained. Filter through Whatman no.41 filter paper and collect the filtrate in a 250mL volumetric flask. Wash the conical flask with small increments of hot water and add the washings to the filter paper.
  - Wash the residue on the filter paper also with hot water and continue the washings till the filtrate runs free of chloride (test with silver nitrate solution).
  - Cool the volumetric flask under tap water and make up the volume to 250mL with cold distilled water. Reserve the triple acid extract so prepared for the analysis of various mineral constituents except N in the given sample.

## 39.6 PROCEDURE

- Pipette out 5mL of the triple acid extract into a 25mL volumetric flask.
- Add 5mL of Barton's reagent, shake well and make up the volume to 25mL with distilled water.
- Allow 30 minutes for the development of yellow color (the color is stable for 24 hours).
- Read the intensity of yellow color developed in a spectrophotometer at 470nm, after adjusting transmittance of the meter to 100 with a blank.
- From the standard curve, deduce the concentration in the solution and from that value, calculate the percentage of total phosphorus content of the plant sample.

### 39.6.1 Preparation of Standard Curve

- Dissolve 0.4390g of pure $KH_2PO_4$ (AR grade) in water and make up the volume to 1000mL with distilled water. This is the stock solution representing 100 ppm.

- Pipette out 10mL of 100 ppm solution into a 100mL volumetric flask and make up to the mark which gives 10ppm.
- 0.5mL of 10ppm diluted to 25mL—0.2ppm

  1.0mL of 10ppm diluted to 25mL—0.4ppm
  1.5mL of 10ppm diluted to 25mL—0.6ppm
  2.0mL of 10ppm diluted to 25mL—0.8ppm
  2.5mL of 10ppm diluted to 25mL—1.0ppm

- After pipetting out a known aliquot of the 10ppm solution into the respective volumetric flask, develop the color as described below:

  - Pipette out 5mL of 10ppm solution.
  - Add 5mL of Barton's reagent and mix well.
  - Make up the volume with water.
  - Measure the intensity of color in a spectrophotometer at 470nm.
  - Plot the readings against concentration to get the standard curve.

## 39.7 OBSERVATION AND CALCULATION

| | | |
|---|---|---|
| Weight of the sample used in the extraction | = | 1g |
| Volume of the triple acid extract made up after digestion | = | V (250) mL |
| Volume of aliquot pipetted out for estimation | = | 5mL |
| Volume made up to | = | 25mL |
| Concentration of P reading in the standard curve corresponding to the meter reading | = | X mg / mL (ppm) |
| ∴ Amount of P present in the plant sample (% P) on moisture free basis | = | $(X / 10^6) \times (25/5) \times V \times (100/1) \times (100/100\text{-}M)$ [M = moisture content of the sample] |

## 39.8 INTERPRETATION

| P content in % of dried material (cereals) | Comments |
|---|---|
| < 0.1 | Deficiencies with symptoms |
| 0.3–0.6 | Optimum supply zone |
| > 1.0 | Toxicity |

## 39.9 THINGS TO LEARN

- How will you prepare one liter of 1ppm, 1*N*, and 1 ***M*** solutions of P using $KH_2PO_4$?
- Why is triple acid extraction done at 180–200°C?
- Give the critical level of P in plants.

# 40 Determination of Total Potassium in Plants *(Flame Photometric Method)*

## 40.1 INTRODUCTION

A plant analysis is used as a diagnostic tool for identifying the plant nutrient status at any stage of crop. It is based on the concept that the concentration of an essential element in a plant or part of the plant indicates the soil's ability to supply that nutrient.

Plant analysis usually refers to the quantitative analysis for the total amount of essential elements in a plant tissue. The potassium (K) content of plants varies from 0.10% to 12%.

## 40.2 IMPORTANCE

- Plant analysis is absolutely necessary in order to examine the nutrition attentively with the view to establishing deficiencies, abundance, or excess.
- From the results of plant analysis, it is possible to make appropriate nutrient management techniques for increasing the crop productivity.

## 40.3 PRINCIPLE

Liquid samples containing certain elements like potassium emit radiation when excited in flame. The excitation causes one of the outer electrons of neutral atoms to jump to an outer orbit of higher energy level or the atoms may be excited sufficiently to lose an electron completely.

When excited atoms return to lower energy levels, light of characteristic wavelength is emitted. The flame photometer measures this radiation intensity, which is proportional to the concentration of elements (like K) in solution. By measuring the intensity in a flame photometer, the K content is determined.

## 40.4 APPARATUS AND MATERIALS REQUIRED

- 25mL volumetric flask
- Pipette
- Beaker
- Funnel

DOI: 10.1201/9781003430100-40

- Balance
- Whatman no. 41 filter paper
- Sand bath
- Flame photometer

## 40.5 REAGENTS REQUIRED

- Ammonium hydroxide
- Red litmus
- Standard K solution
- Triacid extract
  - Weigh 1g of the sample into a 250mL conical flask.
  - Add 12–15mL of triacid mixture (nitric acid:sulfuric acid:perchloric acid at 9:2:1 ratio) and cover the mouth of the flask with a funnel.
  - Digest the contents over a sand bath at 180–200°C until a clear solution is obtained. Filter through Whatman no. 41 filter paper and collect the filtrate in a 250mL volumetric flask.
  - Wash the conical flask with small increments of hot water and add the washings to the filter paper.
  - Wash the residue on the filter paper also with hot water and continue the washings till the filtrate runs free of chloride (test with silver nitrate solution).
  - Cool the volumetric flask under tap water and make up the volume to 250mL with cold distilled water.

## 40.6 PROCEDURE

- Pipette out 5mL of the triple acid extract into a 25mL volumetric flask and neutralize the acid with ammonium hydroxide (the piece of red litmus put into the flask turns to blue).
- Make up the volume with distilled water. Mix well to make the solution homogenous.
- Transfer the content to an injection vial and feed to a flame photometer after adjusting the flame photometer to read 0 with distilled water and 100 with 100ppm K solution.
- Note the meter reading and deduce the concentration of the K solution from the standard curve and calculate the percentage of potassium in the plant sample.

### 40.6.1 Preparation of Standard Curve

- Dissolve 1.907g of KCl in water and make up the volume to 1000mL with distilled water. This stock solution gives 1000ppm of K.
- 100mL of 1000ppm K solution diluted to 1 liter will give 100ppm solution.
- From this, various standards are prepared ranging from 10 to 90ppm as described below:

10.0mL of 100ppm diluted to 100mL—10.0ppm
20.0mL of 100ppm diluted to 100mL—20.0ppm
30.0mL of 100ppm diluted to 100mL—30.0ppm
40.0mL of 100ppm diluted to 100mL—40.0ppm
50.0mL of 100ppm diluted to 100mL—50.0ppm
60.0mL of 100ppm diluted to 100mL—60.0ppm
70.0mL of 100ppm diluted to 100mL—70.0ppm
80.0mL of 100ppm diluted to 100mL—80.0ppm
90.0mL of 100ppm diluted to 100mL—90.0ppm

## 40.7 OBSERVATION AND CALCULATION

| | | |
|---|---|---|
| Weight of the sample used in the extraction | = | 1.0g |
| Volume of the triple acid extract made up after digestion | = | V (250) mL |
| Volume of aliquot pipetted out for estimation | = | 5mL |
| Volume made up to | = | 25mL |
| Concentration of K reading in the standard curve corresponding to the meter reading | = | X mg / mL (ppm) |
| ∴ Amount of K present in the plant sample (% K) on moisture free basis | = | $(X / 10^6) \times (25/5) \times V \times (100/1) \times (100/100\text{-}M)$ [M = moisture content of the sample] |

## 40.8 INTERPRETATION

| K content in % of dried material (cereals) | Comments |
|---|---|
| < 1.2 | Deficiencies with symptoms |
| 2.0–4.0 | Optimum supply zone |
| 5.0 | Toxicity |

## 40.9 THINGS TO LEARN

- How will you prepare one liter of 1ppm, 1*N*, and 1***M*** solutions of P using KCl?
- What are the precautions to be taken while determining the K content of the sample using flame photometer?
- Give the critical level of K in plants.
- Describe the deficiency and toxicity symptoms of K in plants. How will you correct the deficiency and toxicity symptoms?

# 41 Assessment of Quality of Irrigation Water

## 41.1 IMPORTANCE

The quality of irrigation water is a crucial factor for long-term soil productivity. Poor quality water, if used for a long time, can make the soil less productive or even barren, depending on the amount and type of constituents present in it and the texture of the soil in question. Low or marginally saline waters sometimes appear to stimulate crop growth because of the higher amounts of nutrient ions present. However, excess of the soluble salts in water leads to their accumulation in the surface layer particularly in fine-textured or poorly drained soils.

Many areas are facing a serious problem of not only scarcity of water, but also of its extremely poor quality. Tube well or well waters generally pose such problems more than canal waters, especially in arid and semi-arid regions. It is therefore, advisable to get water tested for quality while installing a tube well instead of repenting later. Locating appropriate strata during boring for a tube well is to be done by hit and trial method causing huge investments if the knowledge about surrounding localities is lacking. In such a situation, samples of water should be tested while installation is in progress and the appropriate depth decided accordingly. Irrespective of the source of irrigation water, all that is required to be done is to draw a proper sample for testing in the laboratory.

## 41.2 CRITERIA FOR ASSESSMENT OF QUALITY OF IRRIGATION WATER

### 41.2.1 Salinity Hazard

- Electrical conductivity (EC)
- Total soluble salts (TSS)/total dissolved salts (TDS)
- Osmatic pressure (OP)
- Total cation concentration (TCC)

DOI: 10.1201/9781003430100-41

### 41.2.2 Sodicity Hazard

- Sodium adsorption ratio (SAR)
- Adjusted sodium adsorption ratio (Adj. SAR)
- Sodium to calcium activity ratio (SCAR)
- Sodium ratio
- Figure of merit

### 41.2.3 Salinity and Sodicity Hazard

- Salt index

### 41.2.4 Alkalinity Hazard

- Residual sodium carbonate (RSC)
- Residual sodium bicarbonate (RSBC)

### 41.2.5 Permeability Hazard

- Permeability index

### 41.2.6 Specific Ion Toxicity Hazard

- Sodium hazard–soluble sodium percentage (SSP)
- Magnesium hazard–magnesium adsorption ratio (MAR)
- Chloride hazard
- Sulfate hazard–potential salinity (PS)
- Boron hazard
- Nitrate hazard
- Selenium hazard
- Lithium hazard
- Fluorine hazard

## 41.3 CHEMICAL ANALYSIS OF WATER

### 41.3.1 Determination of Carbonate and Bicarbonate

#### 41.3.1.1 Principle

The carbonate and bicarbonate in the irrigation water are determined by titrating a known volume of water with 0.1*N* $H_2SO_4$ using phenolphthalein and methyl orange indicators.

#### 41.3.1.2 Apparatus and Materials Required

- 250mL conical flask
- 50mL pipette
- Measuring cylinder

- Burette
- Analytical balance

#### 41.3.1.3 Reagents Required

- 0.1 *N* $H_2SO_4$
- Phenolphthalein
- Methyl orange

#### 41.3.1.4 Procedure

- Pipette out 50mL of water sample into a 250mL conical flask.
- Add few drops of phenolphthalein indicator.
- If pink color develops, then proceed to titrate with 0.1 *N* $H_2SO_4$. The end point is disappearance of pink color. This titer value indicates half the amount of carbonate present in the water sample.
- If pink color does not develop, it indicates the absence of carbonate in the water sample.
- Add few drops of methyl orange indicator to the same solution in the conical flask, and observe for the end point wherein the color changes to pinkish red. This titer value indicates half the amount of carbonate plus all the bicarbonate present in the water sample.

#### 41.3.1.5 Observation and Calculation

| | | |
|---|---|---|
| Volume of the water sample taken | = | 50mL |
| Volume of *N*/10 $H_2SO_4$ used for the titration for finding carbonate with phenolphthalein indicator | = | A mL |
| Volume of *N*/10 $H_2SO_4$ used for the titration for finding bicarbonate with methyl orange indicator | = | B mL |
| Actual volume of *N*/10 $H_2SO_4$ used to neutralize carbonate alone | = | (2 × A) mL |
| Actual volume of *N*/10 $H_2SO_4$ used to neutralize bicarbonate alone | = | (B–A) mL |
| 1mL of *N*/10 $H_2SO_4$ | = | 0.1meq of $CO_3$ |
| ∴ (2 × A) mL of *N*/10 $H_2SO_4$ | = | 0.2 × A meq of $CO_3$ |
| This is present in 50mL of the water sample | | |
| ∴ the concentration of carbonate in the water sample | = | 0.2 × A (1000/50) meq / lit of $CO_3$ |
| 1 mL of *N*/10 $H_2SO_4$ | = | 0.1meq of $HCO_3$ |
| ∴ (B–A) mL of *N*/10 $H_2SO_4$ | = | 0.1 × (B–A) meq of $HCO_3$ |
| This is present in 50mL of the water sample | | |
| ∴ the concentration of bicarbonate in the water sample | = | 0.1 × (B–A) × (1000/50) meq / lit of $HCO_3$ |

### 41.3.2 Determination of Chloride

#### 41.3.2.1 Principle

Chloride present in the water sample is precipitated as silver chloride by titrating it with 0.1 *N* $AgNO_3$ (Mohr's titration) in the presence of potassium chromate indicator.

An additional drop of silver nitrate added after the complete precipitation of chloride will combine with potassium chromate forming flesh/brick red colored silver chromate precipitate (end point).

#### 41.3.2.2 Apparatus and Materials Required

- Porcelain basin
- Burette
- 50mL pipette
- Analytical balance

#### 41.3.2.3 Reagents Required

- 0.1 *N* $AgNO_3$
- 5% aqueous potassium chromate indicator

#### 41.3.2.4 Procedure

- Pipette out 50mL of water sample into a porcelain basin.
- Add few drops of potassium chromate indicator.
- Titrate it against 0.1 *N* $AgNO_3$ till a flesh/brick red precipitate of silver chromate is formed and note down the titer value.
- From the titer value, calculate the chloride present in the water sample.

#### 41.3.2.5 Observation and Calculation

| | | |
|---|---|---|
| Volume of the water sample taken | = | 50mL |
| Volume of *N*/10 $AgNO_3$ used for the titration | = | A mL |
| 1mL of *N*/10 $AgNO_3$ | = | 0.1meq of $Cl^-$ |
| ∴ A mL of *N*/10 $AgNO_3$ | = | 0.1 × A meq of $Cl^-$ |
| This is present in 50mL of the water sample | | |
| ∴ the concentration of chloride in the water sample | = | 0.1 × A × (1000/50) meq / lit of $Cl^-$ |

### 41.3.3 Determination of Calcium and Magnesium

#### 41.3.3.1 Principle

The EDTA (ethylene diamine tetra acetic acid) forms stable complexes with various polyvalent cations at different pH level. Among calcium and magnesium, calcium is complexed by EDTA first followed by magnesium at pH 12 and 10 respectively.

The calcium in the sample is estimated by titrated it with EDTA using the indicator ammonium purpurate, commonly known as "murexide," in the presence of sodium hydroxide buffer at pH 12.

The calcium plus magnesium is determined by titration with EDTA using Eriochrome black-T (EBT) as indicator, in the presence of ammonium chloride–ammonium hydroxide buffer at pH 10.

#### 41.3.3.2 Apparatus and Materials Required

- Porcelain basin
- Burette
- 50mL pipette
- Analytical balance

#### 41.3.3.3 Reagents Required

- 0.02*N* EDTA
- 10% NaOH buffer
- Murexide (ammonium purpurate) indicator
- Ammonium chloride–ammonium hydroxide buffer
- Eriochrome black-T indicator (EBT indicator)

#### 41.3.3.4 Procedure

- Calcium alone
  - Pipette out 25mL of water sample into a porcelain basin.
  - Add 10mL of 10% NaOH buffer and 50mg of murexide indicator.
  - Titrate it against 0.02*N* EDTA till the color changes from pink red to purple or violet.
  - From the titer value, calculate the concentration of calcium present in the water sample.
- Calcium plus magnesium
  - Pipette out 25mL of water sample into a porcelain basin.
  - Add 10mL of ammonium chloride–ammonium hydroxide buffer solution.
  - Add 4–5 drops of EBT indicator.
  - Titrate it against 0.02*N* EDTA till the color changes from wine red to sky blue.
  - From the titer value, calculate the concentration of magnesium present in the water sample.

#### 41.3.3.5 Observation and Calculation

| | | |
|---|---|---|
| Volume of the water sample taken | = | 25mL |
| Volume of 0.02*N* EDTA used in the estimation of Ca + Mg (EBT indicator) | = | A mL |
| Volume of 0.02*N* EDTA used in the estimation of Ca alone (murexide indicator) | = | B mL |
| Volume of 0.02*N* EDTA used in the estimation of Mg alone | = | (A–B) mL |
| 1mL of 0.02*N* EDTA | = | 0.02meq of $Ca^{2+}$ |
| ∴ B mL of 0.02*N* EDTA | = | 0.02 × B meq of $Ca^{2+}$ |
| This is present in 25mL of the water sample | | |
| ∴ the concentration of calcium in the water sample | = | 0.02 × B × (1000/25) meq / lit of $Cl^-$ |
| 1mL of 0.02*N* EDTA | = | 0.02meq of $Mg^{2+}$ |
| ∴ (A–B) mL of 0.02 *N* EDTA | = | 0.02 × (A–B) meq of $Mg^{2+}$ |
| This is present in 25mL of the water sample | | |
| ∴ the concentration of calcium in the water sample | = | 0.02 × (A–B) × (1000/25) meq / lit of $Mg^{2+}$ |

## 41.3.4 Determination of Sodium and Potassium

### 41.3.4.1 Principle

Sodium and potassium in the water sample is estimated by using flame photometry. The samples fed are atomized over a flame and the desired ions are excited in it. The intensity of the light emitted is proportional to the concentration of the desired ions in the given sample.

### 41.3.4.2 Apparatus and Materials Required

- Flame photometer
- Volumetric flasks
- Beaker
- Pipette
- Analytical balance

### 41.3.4.3 Reagents Required

- AR grade KCl
- AR grade NaCl
- 100ppm stock solution of K (0.1907g of AR grade KCl in 1 liter of distilled water)
- Working standard solutions of K (2, 4, 6, 8, 10, and 20ppm of K)
- 100ppm stock solution of Na (0.254g of AR grade NaCl in 1 liter of distilled water)
- Working standard solutions of Na (5, 10, 15, 20, 30, 40, and 50ppm of Na)

### 41.3.4.4 Procedure

Potassium

- Filter a portion of the water sample if suspended material is visible.
- Filtration is desirable as it prevents choking of capillary tube of the flame photometer.
- Take the working standards solutions and record the flame photometer reading against each, after setting zero with distilled water and 100 with the highest concentration of K (20ppm of K).
- Feed the test sample and record the reading. Draw a standard curve by plotting the flame photometer reading against K concentration.
- Find out the concentration of K from standard graph and calculate the concentration of K in the given water sample.

Sodium

- Filter a portion of the water sample if suspended material is visible.
- Filtration is desirable as it prevents choking of capillary tube of the flame photometer.

- Take the working standards solutions and record the flame photometer reading against each, after setting zero with distilled water and 100 with the highest concentration of Na (50ppm of Na).
- Feed the test sample and record the reading. Draw a standard curve by plotting the flame photometer reading against Na concentration.
- Find out the concentration of Na from standard graph and calculate the concentration of Na in the given water sample.

Note: It is also possible to find out the flame photometer reading for Na and K simultaneously, if the desired make and model of flame photometer is available in the laboratory.

### 41.3.4.5 Observation and Calculation

| | | |
|---|---|---|
| The concentration of potassium derived from the standard graph in the water sample | = | A ppm |
| ∴ the concentration of potassium in the water sample | = | (A / 39) meq / lit of K |
| The concentration of sodium derived from the standard graph in the water sample | = | B ppm |
| ∴ the concentration of potassium in the water sample | = | (B / 23) meq / lit of Na |

Note: If the flame photometer reading exceeds 100 for the test sample, dilute the test same (for example, 5mL test sample may be diluted to 25mL with distilled water) and feed the same. If diluted, the dilution factor (for the given example, the dilution factor is (25/5)) is to be included in the calculation accordingly.

# Bibliography

Baruah, T.C. & Barthakur, H.P. *A Text Book of Soil Analysis*. Vikas Publishing House Pvt. Ltd., New Delhi, 1998.

Black, C.A. *Methods of Soil Analysis*. American Society of Agronomy, Inc., Madison, WI, 1965.

Bray, R.H. & Kurtz, L.T. Determination of total, organic and available forms of phosphorus in soils. *Soil Science*, 1945, 59: 39–45.

Chopra, S.L. & Kanwar, J.S. *Analytical Agricultural Chemistry*. Kalyani Publishers, New Delhi, 1986.

Dickman, S.R. & Bray, R.H. Colorimetric determination of phosphate. *Industrial and Engineering Chemistry Analytical Edition*, 1940, 12: 665–668.

Durai, M.V. *Handbook of Soil, Plant, Water, Fertilizers and Manure Analysis*. New India Publishing Agency, New Delhi, 2014.

Ghosh, A.B., Bajaj, J.C., Hasan, R. and Singh, D. *Soil and Water Testing Methods: A Laboratory Manual*. Indian Agricultural Research Institute, New Delhi, 1983.

Gupta, P.K. *Soil, Plant, Water and Fertilizer Analysis*. Agrobios (India), Jodhpur, 2007.

Hanway, J.J. and Heidel, H. Soil analysis methods as used in Iowa State College Soil Testing Laboratory. *Iowa State College of Agriculture Bulletin*, 1952, 57: 1–31.

Hesse, P.R. *A Textbook of Soil Chemical Analysis*. CBS Publishers and Distributors, New Delhi, 1994.

Jackson, M.L. *Soil Chemical Analysis*. Prentice Hall, New Delhi, 1973.

Jaiswal, P.C. *Soil, Plant and Water Analysis*. Kalyani Publishers, Ludhiana, 2011.

Lindsay, W.L. & Norvell, W.A. Development of a DTPA soil test for zinc, iron, manganese and copper. *Soil Science Society of America Journal*, 1978, 42: 421–448.

Mailappa, A.S. *Analytical Techniques and Instrumental Methods in Soil and Plant Analysis*. Brilliant Publishing, New Delhi, 2020.

Mailappa, A.S. *Soil Fertility and Plant Nutrition Manual*. Brilliant Publishing, New Delhi, 2021.

OECD. *Principles of Good Laboratory Practice*. Paris, Organisation for Economic Co-operation and Development, 1998.

Olsen, S.R., Cole, C.V., Watanable, F.S. & Dean, L.A. Estimation of available phosphorus in soils by extraction with sodium bicarbonate. *Circular: US Department of Agriculture*, 1954, 939 : 1–19.

Oregon State University. *Health and Safety Training Manual*, Corvallis, Oregon. Oregon State University.

Page, A.L., Miller, R.H. & Keeney, D.R. Methods of Soil Analysis, Part I1, Chemical and Microbiological Properties, Agron. Monograph No. 9. Madison, WI, American Society of Agronomy, 1982.

Piper, C.S. *Soil and Plant Analysis*. Bombay, Hans Publishers, 1966.

Ratan, R.K., Katyal, J.C., Dwivedi, B.S., Sarkar, A.K., Bhattacharyya, T., Tarafdar, J.C. & Kukal, S.S. *Soil Science: An Introduction*. New Delhi, Indian Society of Soil Science, 2015.

Robinson, J.W., Skelly Frame, E.M. & Frame II, G.M. *Undergraduate Instrumental Analysis*. Boco Raton, FL, CRC Press, 2014.

Schoonover, W.R. *Examination of Soils for Alkali*. Berkeley, CA, University of California, 1952.

Schwarzenbach, G., Kampitsch, E. & Steiner, R. Komplexone III: Uramil-diessigsäure und ihr Komplexbildunsvermögen. *Helvetica Chimica Acta*, 1946, 29(2): 364–370.

Shoemaker, H.E., McLean, E.O. & Pratt, P.F. Buffer methods for determining lime requirement of soils with appreciable amounts of extractable aluminum. *Soil Science Society of America Journal*, 1961, 25: 274–277.

Singh, D., Chhonkar, P.K. & Dwivedi, B.S. *Manual on Soil, Plant and Water Analysis*. New Delhi, Westville Publishing House, 2007.

Singh, D., Chhonkar, P.K. & Pandey, R.N. *Soil Plant Water Analysis: A Methods Manual*. New Delhi, Indian Agricultural Research Institute, 1999.

Smith, K.A. *Soil Analysis: Modern Instrumental Techniques*. New York, Marcel Dekker, Inc., 1990.

Subbiah, B.V. & Asija, G.L. A rapid procedure for the determination of available nitrogen in soils. *Current Science*, 1956, 25: 259–260.

Tan, K.H. *Soil Sampling, Preparation and Analysis*. Boca Raton, FL, CRC Press, Inc., 2005.

Tandon, H.L.S. *Methods of Analysis of Soils, Plants, Waters and Fertilizers*. New Delhi, FDCO, 2013.

Tisdale, S.L., Nelson, W.L. & Beaton, J.D. *Soil Fertility and Fertilizers*. New York, Macmillan Publishing Company, 1985.

Walkley, A. & Black, I.A. An examination of the Degtjareff method for determining soil organic matter, and a proposed modification of the chromic acid titration method. *Soil Science*, 1934, 34: 29–38.

Willard, H.H., Merrit Jr., L.L. & Dean, J.A. *Instrumental Methods of Analysis*. New Delhi, Affiliated East-West Press Pvt. Ltd., 1974.

Williams, C.H. and Steinbergs, A. Soil sulfur fractions as chemical indices of available sulfur in some Australian soils. *Australian Journal of Agricultural Research*, 1959, 10: 340–352.

# Appendices

## APPENDIX I
## MOLECULAR AND EQUIVALENT WEIGHTS OF SOME IMPORTANT COMPOUNDS

| Name | Formula | Mol. wt (g) | Eq. wt (g) |
|---|---|---|---|
| Ammonium acetate | $CH_3COONH_4$ | 77.08 | 77.08 |
| Ammonium chloride | $NH_4Cl$ | 53.49 | 53.49 |
| Ammonium fluoride | $NH_4F$ | 37.04 | 37.04 |
| Ammonium nitrate | $NH_4NO_3$ | 80.04 | 80.04 |
| Barium acetate | $(CH_3COO)_2Ba$ | 255.43 | 127.72 |
| Barium chloride | $BaCl_2.2H_2O$ | 244.28 | 122.14 |
| Boric acid | $H_3BO_3$ | 61.83 | 20.61 |
| Calcium acetate | $(CH_3COO)_2Ca$ | 158.00 | 79.00 |
| Calcium carbonate | $CaCO_3$ | 100.09 | 50.05 |
| Calcium chloride (dihydrate) | $CaCl_2.2H_2O$ | 147.02 | 73.51 |
| Calcium hydroxide | $Ca(OH)_2$ | 74.00 | 37.00 |
| Calcium nitrate | $Ca(NO_3)_2$ | 164.00 | 82.00 |
| Calcium sulfate | $CaSO_4.\ 2H_2O$ | 172.17 | 86.08 |
| Ferrous ammonium sulfate | $NH_4SO_4.FeSO_4.6H_2O$ | 392.13 | 392.13 |
| Ferrous sulfate | $FeSO_4.7H_2O$ | 278.01 | 139.00 |
| Magnesium chloride | $MgCl_2.6H_2O$ | 203.30 | 101.65 |
| Magnesium nitrate | $Mg(NO_3)_2.6H_2O$ | 256.41 | 128.20 |
| Potassium chloride | KCl | 74.55 | 74.55 |
| Potassium dichromate | $K_2Cr_2O_7$ | 294.19 | 49.04 |
| Potassium hydroxide | KOH | 56.10 | 56.10 |
| Potassium permanganate | $KMnO_4$ | 158.03 | 31.60 |
| Potassium nitrate | $KNO_3$ | 101.10 | 101.10 |
| Potassium sulfate | $K_2SO_4$ | 174.27 | 87.13 |
| Potassium hydrogen phthalate | $COOH\ C_6H_4COOK$ | 204.22 | 204.22 |
| Oxalic acid | $C_2H_2O_4.2H_2O$ | 126.00 | 63.00 |
| Silver nitrate | $AgNO_3$ | 169.87 | 169.87 |
| Sodium acetate | $CH_3COONa$ | 82.04 | 82.04 |
| Sodium bicarbonate | $NaHCO_3$ | 84.01 | 84.01 |
| Sodium carbonate | $Na_2CO_3$ | 106.00 | 53.00 |
| Sodium chloride | NaCl | 58.45 | 58.45 |
| Sodium hydroxide | NaOH | 40.00 | 40.00 |
| Sodium nitrate | $NaNO_3$ | 84.99 | 84.99 |
| Sodium oxalate | $Na_2C_2O_4$ | 134.00 | 67.00 |
| Sodium sulfate | $Na_2SO_4$ | 142.04 | 71.02 |
| Sodium thiosulfate | $Na_2S_2O_3.5H_2O$ | 248.18 | 248.18 |

# APPENDIX II
# GUIDELINES FOR THE PREPARATION OF STANDARD SOLUTION

| Sl. No. | To prepare 1 liter of 0.1*N* solution (approximately) | Standardize with |
|---|---|---|
| 1. | Sulfuric acid ($H_2SO_4$) (Dilute 3.0mL of concentrated 36*N* acid to 1 liter with water) | *N*/10 solution of $Na_2CO_3$ (5.29g of AR grade dry $Na_2CO_3$ in water and made up to 1 liter) (or) *N*/10 borax solution [Dissolve 1.907 g of A.R. grade borax ($Na_2B_4O_7.10H_2O$) in water and make up to 100 mL] |
| 2. | Hydrochloric acid (HCl) (Dilute 10mL of concentrated 12*N* acid to 1 liter with water) | |
| 3. | Nitric acid ($HNO_3$) (Dilute 6.3mL of concentrated 16*N* acid to 1 liter with water) | |
| 4. | Sodium hydroxide (NaOH) (Dissolve 4.0g in water and make up to 1 liter) | *N*/10 solution of potassium hydrogen phthalate ($HKC_8H_4O_4$) (Dissolve 5.1g of AR potassium hydrogen phthalate in water and make up to 250mL) (or) *N*/10 solution of succinic acid (Dissolve 5.0g of succinic acid in water and make up to 1 liter) |
| 5. | Potassium hydroxide (KOH) (Dissolve 5.7–6.0g in water and make up to 1 liter) | |
| 6. | Silver nitrate ($AgNO_3$) (Dissolve 16.989g crystalline AR silver nitrate in water and make up to 1 liter) | *N*/10 solution of NaCl (Dissolve 5.847g AR sodium chloride in water and make up to 1 liter) |
| 7. | Potassium dichromate ($K_2Cr_2O_7$) (Dissolve 4.904g AR potassium dichromate in water and make up to 1 liter) | Potassium dichromate may be used as a primary standard |
| 8. | Potassium permanganate ($KMnO_4$) (Dissolve 3.25 g potassium permanganate in water and make up to one liter) | *N*/10 Oxalic acid (Dissolve 1.574g pure oxalic acid [$(COOH)_2$. 2 $H_2O$] in water and make up to 250mL) (or) *N*/10 sodium oxalate (Dissolve 1.67g pure sodium oxalate in water and make up to 250mL) |

| Sl. No. | To prepare 1 liter of 0.1*N* solution (approximately) | Standardize with |
|---|---|---|
| 9. | Sodium thiosulfate ($Na_2S_2O_3$) (Dissolve 25.0 g pure sodium thiosulfate in boiled out distilled water and make up to 1 liter. Add 3.8g of borax or 0.1g of anhydrous sodium carbonate to stabilize the solution) | *N*/10 $KMnO_4$ solution (or) *N*/10 potassium iodate ($KIO_3$) (or) *N*/10 iodine |
| 10. | Iodine (I) (Dissolve 20.0 g AR iodate free potassium iodide in 30–40mL of water and add 12.7g sublimed Iodine and make up to 1 liter) | *N*/10 sodium thiosulfate ($Na_2S_2O_3$) |
| 11. | Fehling's solution A (Dissolve 69.28 g pure copper sulfate ($CuSO_4$) in water and make up to 1 liter) Fehling's solution B (Dissolve 350.0g Rochelle salt [potassium sodium tartarate] and 100g of sodium hydroxide in water and make up to 1 liter) | 0.5% solution of pure AR glucose (5mL of Fehling's solution A and 5mL of Fehling's solution B to titrate against 10mL of the above standard solution of dextrose) |
| 12. | Potassium thiocyanate (KCNS) (Dissolve 10g potassium thiocyanate in water and make up to 1 liter) | *N*/10 silver nitrate |
| 13. | EDTA solution (Dissolve 20g sodium salt in water and make up to 1 liter) | *N*/10 calcium solution |
| 14. | Ferrous sulfate ($Fe_2SO_4.7\ H_2O$) (Dissolve 27.8g in water and make up to 1 liter) | *N*/10 potassium dichromate |
| 15. | Ferrous ammonium sulfate ($Fe_2SO_4\ (NH_4)_2SO_4.7\ H_2O$) (Dissolve 39.2g ferrous ammonium sulfate in water and make up to 1 liter) | *N*/10 potassium dichromate |

# APPENDIX III
# STRENGTH OF AQUEOUS SOLUTIONS OF SOME ACIDS AND AQUEOUS AMMONIA

| Sl. No. | Name | Approximate Specific gravity | Approximate Percentage by weight | Approximate Normality | Volume required to make 1 liter of approximately 0.1 *N* solution (mL) |
|---|---|---|---|---|---|
| 1. | HCl | 1.19 | 39 | 11.8 | 7.8 |
| 2. | $HNO_3$ | 1.42 | 71 | 15.6 | 6.2 |
| 3. | $H_2SO_4$ | 1.84 | 96 | 36.0 | 2.8 |
| 4. | $H_3PO_4$ | 1.70 | 85 | 44.0 | 2.3 |
| 5. | $CH_3COOH$ | 1.06 | 99.0 | 17.5 | 5.8 |
| 6. | Aqueous $NH_3$ | 0.91 | 28.3 ($NH_3$) | 15.0 | 6.7 |
| 7. | $H_2O$ | 1.00 | 100 | - | - |

# APPENDIX IV
# CHOICE OF INDICATORS

| Sl. No. | Type of titration | Example of reactions | Suitable indicator | pH range of color change |
|---|---|---|---|---|
| 1. | Strong acid vs strong base | NaOH Vs HCl<br>NaOH Vs $H_2SO_4$<br>KOH Vs HCL etc., | Methyl orange<br>Methyl red<br>Bromethyl blue<br>Phenolphthalein | 5.0 to 10.0 |
| 2. | Strong acid vs weak base | HCL Vs $Na_2CO_3$<br>HCl Vs $NaHCO_3$<br>HCl Vs $Ca(OH)_2$<br>$H_2SO_4$ Vs $Na_2CO_3$ etc., | Methyl orange<br>Methyl red | 4.0 to 6.0 |
| 3. | Weak acid vs strong base | $CH_3COOH$ Vs NaOH<br>$CH_3COOH$ Vs KOH | Phenolphthalein or thymolphthalein | 8.0 to 10.0 |
| 4. | Weak acid vs weak base | $CH_3COOH$ Vs $Na_2CO_3$<br>$CH_3COOH$ Vs $NaHCO_3$ | No suitable indicator (potentiometric titration can be used) | Slow change of pH |

# APPENDIX V

| Sl. No. | Name of the indicator | Color change | | |
|---|---|---|---|---|
| | | Acid | Base | pH range |
| **A. Acid-base indicators** | | | | |
| 1. | Methyl violet | Yellow | Blue | 0.5–1.5 |
| 2. | Thymol blue | Red | Yellow | 1.2–2.8 |
| 3. | Methyl yellow | Red | Yellow | 2.9–4.0 |
| 4. | Methyl orange | Red | Yellow | 3.1–4.5 |
| 5. | Bromophenol blue | Yellow | Bluish violet | 30.–4.6 |
| 6. | Bromocresol green | Yellow | Blue | 3.6–5.2 |
| 7. | Methyl red | Red | Yellow | 4.2–6.3 |
| 8. | Bromothymol blue | Yellow | Blue | 6.0–7.6 |
| 9. | Paranitro phenol | Colorless | Yellow | 6.2–7.5 |
| 10. | Litmus | Red | Blue | 5.0–8.0 |
| 11. | Phenol red | Yellow | Red | 6.4–8.2 |
| 12. | Cresol red | Yellow | Red | 7.2–8.8 |
| 13. | Thymol blue | Yellow | Blue | 8.0–9.6 |
| 14. | Phenolphthalein | Colorless | Pinkish red to pink | 8.0–10.0 |
| 15. | Thymolphthalein | Colorless | Blue | 9.3–10.5 |

**B. Redox indicators**

| Sl.No. | Name of the indicator | In (ox) | In (red) | E at $H^+ = 1$ (volt) |
|---|---|---|---|---|
| 16. | Methylene blue | Blue | Colorless | + 0.53 |
| 17. | Diphenylamine | Blue–violet | Colorless | + 0.76 |
| 18. | Ferroin ($Fe^+$ ortho phenolphthalein) | Pale blue | Red | + 1.14 |
| 19. | Neutral red | Red | Colorless | + 0.24 |

**C. Miscellaneous**

| Sl. No. | Name of the indicator | Color change |
|---|---|---|
| 20. | Potassium chromate | White to flesh red |
| 21. | Starch | Colorless to blue/blue to colorless |
| 22. | Potassium ferricyanide | No blue color/colorless to brown |
| 23. | Ammonium purpurate (murexide) | Colorless to brown |
| 24. | Eriochrome black-T | Purple to green or sky blue |

# APPENDIX VI
# CONVERSION FACTORS

| Non-SI units | Conversion factor | SI units |
|---|---|---|
| **Multiply this** | **by** | **to obtain this** |
| | **Length** | |
| Angstrom, A° ($10^{-10}$ m) | 0.1 | nanometer, nm ($10^{-9}$ m) |
| foot, ft | 0.304 | meter, m |
| inch, in | 25.40 | millimeter, mm ($10^{-3}$ m) |
| micron, μ | 1.00 | micrometer, μm ($10^{-6}$ m) |
| mile | 1.609 | kilometer, km ($10^{3}$ m) |
| | **Mass** | |
| ounce, oz | 28.350 | grams, g ($10^{-3}$ kg) |
| pound, lb | 453.592 | grams, g ($10^{-3}$ kg) |
| quintal (metric), q | 100 | kilogram, kg |
| tonne/ton (metric), t | 1000 | kilogram, kg |
| ton (2000lb), t | 907.185 | kilogram, kg |
| | **Volume** | |
| gallon (US), gal | 3.785 | liter, L ($10^{-3}$ $m^3$) |
| ounce, fluid, oz | 0.030 | liter, L ($10^{-3}$ $m^3$) |
| cubic foot, $ft^3$ | 0.028 | cubic meter, $m^3$ |
| cubic foot, $ft^3$ | 28.317 | liter, L ($10^{-3}$ $m^3$) |
| | **Area** | |
| acre, ac | 0.405 | hectare, ha ($10^{4}$ $m^2$) |
| acre, ac | 4046.85 | square meter, $m^2$ |
| | **Others** | |
| atmosphere | 1.013 | bar, bar |
| bar | 100 | kilopascal, k Pa ($10^{3}$ Pa) |
| calorie, C | 4.19 | joule, J |
| Dyne | $10^{-5}$ | newton, N |
| Erg | $10^{-7}$ | joule, J |
| millimhos/cm, mmhos/cm | 1.00 | deci Siemen/m, $dSm^{-1}$ |
| micromhos/cm, μ mhos/cm | 0.10 | milli Siemen/m, $mSm^{-1}$ |
| pound per acre, lb/a | 1.121 | kilogram per hectare, kg $ha^{-1}$ |
| pounds per square foot, lb/$ft^2$ | 47.88 | pascal, Pa |
| degree Fahrenheit (°F –32) | 0.556 | degree Celsius, °C |
| degree Celsius (°C +273) | 1.00 | degree Kelvin, K |
| water potential (mass basis) in bars | 100 | Joules per kilogram, j $kg^{-1}$ |
| tons (2000 lb) per acre, t/a | 2.242 | Metric tons (1000 kg) per hectare, t $ha^{-1}$ (or) Mega grams per hectare, Mg $ha^{-1}$ |
| millimho per centimeter, mmho $cm^{-1}$ | 0.1 | Siemen per metre, S $m^{-1}$ |

| Non-SI units | Conversion factor | SI units |
|---|---|---|
| | **Special Conversions** | |
| $P_2O_5$ | 0.44 | P |
| $K_2O$ | 0.83 | K |
| MgO | 0.602 | Mg |
| CaO | 0.715 | Ca |
| P | 2.29 | $P_2O_5$ |
| K | 1.20 | $K_2O$ |
| Ca | 1.39 | CaO |
| Mg | 1.66 | MgO |
| ppm | 2.24 | kg/ha |

* To convert from SI to non-SI units, divide by the conversion factor given

## APPENDIX VII
## SOME IMPORTANT UNITS AND RELATIONSHIPS

| | | |
|---|---|---|
| 1 acre | = | $4046.8m^2$ |
| 1 hectare | = | 2.471 acres / $10000m^2$ |
| 1 acre–inch | = | $3630ft^3$ |
| 1 gram | = | $10^6$ micrograms / 100 centigrams / 10 decigrams / 1000 milligrams |
| 1 parts per million (ppm) | = | mg $liter^{-1}$ / microgram $mL^{-1}$ / mg/kg |
| 1% | = | 10g/kg |
| 1% | = | 10000ppm |
| milli equivalents | = | weight in mg / equivalent weight |
| 1 Siemen | = | 1mho |
| 1 mmhos $cm^{-1}$ | = | $1dSm^{-1}$ or $0.1Sm^{-1}$ |
| 1 bar | = | 0.9869 atmosphere |
| 1 atmosphere | = | 33.90ft of water / 76.0cm of mercury |
| 1 tonne / 1 Metric ton | = | 1000kg / 1 Megagram (Mg) |
| 1meq per 100 g soil | = | 1 centimole per kg of soil |
| 1meq ($K^+$) per 100g soil | = | 10mmol ($K^+$) per kg soil |
| 1meq $Ca^{2+}$ per 100g soil | = | 10mmol (1/2 $Ca^{2+}$) per kg soil / 10$mmol_c$ ($Ca^{2+}$) per kg soil |
| 1meq $Al^{3+}$ per 100g soil | = | 10mmol (1/3 $Al^{3+}$) per kg soil / 10$mmol_c$ ($Al^{3+}$) per kg soil |
| 1 CEC (meq per 100g soil) | = | 1 CEC (mmols (+) per 100 g soil) |
| 0.1*M* HCl | = | 0.1mol per liter HCl |
| 1 equivalent of $H_2SO_4$ | = | 1 mole of ½ $H_2SO_4$ / 1 $mole_c$ of $H_2SO_4$ |

## APPENDIX VIII
## SOME IMPORTANT CONVERSION FACTORS

- N × 1.286 = $NH_4$
- N × 4.43 = $NO_3$
- Organic carbon × 1.724 = organic matter
- P × 2.29 = $P_2O_5$
- K × 1.20 = $K_2O$
- Ca × 1.40 = CaO
- Ca × 1.85 = $Ca(OH)_2$
- S × 3 = $SO_4$
- Acre × 2.471 = Hectare = 2.24 × $10^6$ kg soil (plough layer)
- Hectare × 0.404686 = Acre
- 1 pura (local name in Arunachal) = 5184$m^2$ = 1.296 acre = 0.5184 hectare
- Lb / acre × 1.21 = kg / ha
- Meq / L × equivalent weight = ppm
- ppm = μg / mL = mg / L
- ppm = mg / kg
- mm per opening = 16/ mesh per inch

## APPENDIX IX
## SIEVE SIZE

| Opening diameter (mm) | Mesh/inch (approx.) |
|---|---|
| 6 | 4 |
| 4 | 6 |
| 2 | 10 |
| 1 | 20 |
| 0.25 | 60 |
| 0.16 | 100 |
| 0.1 | 140 |

## APPENDIX X
## SOME PREFIX, SYMBOLS, AND THEIR MEANINGS

| Prefix | Meaning | Symbol | Prefix | Meaning | Symbol |
|---|---|---|---|---|---|
| deca | 10 | da | deci | $10^{-1}$ | d |
| hecta | $10^2$ | h | centi | $10^{-2}$ | c |
| kilo | $10^3$ | k | milli | $10^{-3}$ | m |
| mega | $10^6$ | M | micro | $10^{-6}$ | μ |
| giga | $10^9$ | G | nano | $10^{-9}$ | n |
| tera | $10^{12}$ | T | pico | $10^{-12}$ | p |
| peta | $10^{15}$ | P | femto | $10^{-15}$ | f |
| exa | $10^{18}$ | E | atto | $10^{-18}$ | a |
| zetta | $10^{21}$ | Z | zapto | $10^{-21}$ | z |
| yotta | $10^{24}$ | Y | yocto | $10^{-24}$ | y |

# APPENDIX XI
# FERTILITY RATING CHART FOR AVAILABLE MACRONUTRIENTS IN SOILS

| | Rating for medium category | | |
|---|---|---|---|
| **Macronutrients** | **Low** | **Medium** | **High** |
| Available N (kg / ha) | < 280 | 280–560 | > 560 |
| Available P (kg / ha) | < 10 | 10–25 (Olsen P) | > 25 |
| | < 16 | 16–45 (Bray P) | > 45 |
| Available K (kg / ha) | < 118 | 118–280 | > 280 |
| Available/exchangeable Ca (meq / 100g of soil) | < 10 or (< 25% of CEC–deficient) | 10–30 | > 30 |
| Available/exchangeable Mg (meq / 100g of soil) | < 5 or (< 4% of CEC–Deficient) | 5–10 | > 10 |
| Available S (ppm) (0.15% $CaCl_2$ extractable) | < 10 (< 10–deficient) | 10–15 | > 15 |

# APPENDIX XII
# CRITICAL LIMITS/LEVEL OF AVAILABLE MICRONUTRIENTS IN SOILS

| Micronutrients | Critical limits (ppm) |
|---|---|
| DTPA Fe | 4.5 |
| DTPA Mn | 2.0–3.0 |
| DTPA Zn | 0.8 |
| DTPA Cu | 0.2 |
| Hot water-soluble B | 0.5–1.0 |
| Ammonium oxalate extractable / Grigg–Mo | 0.05–0.2 |

# APPENDIX XIII
# AVERAGE NUTRIENT CONTENT (%) OF ORGANIC SOURCES

| Name | N | $P_2O_5$ | $K_2O$ |
|---|---|---|---|
| Farmyard manure | 0.85 | 0.25 | 0.65 |
| Ground nut cake | 7.30 | 1.50 | 1.30 |
| Sesamum cake | 6.20 | 2.00 | 1.20 |
| Neem cake | 5.20 | 1.10 | 1.50 |
| Castor cake | 4.40 | 1.10 | 1.40 |
| Cotton seed cake | 4.00 | 1.90 | 1.60 |
| Coconut cake | 3.00 | 1.90 | 1.80 |
| Sewage sludge | 1.0–6.0 | 0.75–2.0 | 0.40 |
| Dhaincha | 3.50 | 0.60 | 1.20 |
| *Sesbania rostrata* | 2.80 | 0.10 | 1.40 |
| Sunhemp | 2.30 | 0.50 | 1.80 |
| Composted coir pith | 1.06 | 0.40 | 1.20 |
| Fish meal | 4.00 | 3.90 | 1.80 |

| Name | N | $P_2O_5$ | $K_2O$ |
|---|---|---|---|
| Blood meal | 10.10 | 1.20 | 1.00 |
| Guano | 11.00 | 10.00 | 2.50 |
| Night soil compost | 1.40 | 2.80 | 4.00 |
| Goat manure | 2.40 | 0.90 | 2.00 |
| Sheep manure | 1.93 | 1.70 | 2.30 |
| Pig manure | 3.70 | 3.30 | 0.40 |
| Poultry manure | 2.17 | 2.00 | 4.20 |
| Biogas slurry | 1.40 | 0.90 | 0.80 |

## APPENDIX XIV
## NUTRIENT CONTENT, MOISTURE, FREE ACIDITY, AND EQUIVALENT ACIDITY/BASICITY OF NITROGENOUS FERTILIZERS

| | | N Content | | | | | | | |
|---|---|---|---|---|---|---|---|---|---|
| **Fertilizer** | **Formula** | **Total** | **$NH_4$ - N** | **$NO_3$ – N** | **$NH_2$ - N** | **Moisture (%)** | **Free Acidity (%)** | **Equivalent acidity/ basicity** | **Remarks** |
| Ammonium sulfate | $(NH_4)_2 SO_4$ | 20.6 | 20.6 | - | - | 1.0 | 0.025 | 110 | Sulfur–24% |
| Ammonium sulfate nitrate | $(NH_4)_2SO_4 + NH_4NO_3$ | 26.0 | 19.25 | 6.75 | - | 1.0 | 0.015 (as $HNO_3$) | 93 | Sulfur–15% |
| Ammonium chloride | $NH_4Cl$ | 25.0 | 25.0 | - | - | 2.0 | - | 128–140 | Ca as NaCl–2% |
| Anhydrous ammonia | $NH_3$ | 81.5 | 81.5 | - | - | 1.0 | - | 148 | - |
| Calcium ammonium Nitrate | $NH_4NO_3 + CaCO_3$ | 25.0 | 12.5 | 12.5 | - | 1.0 | - | Neutral | - |
| Urea | $CO(NH_2)_2$ | 46.0 | - | - | 46.0 | 1.0 | - | 80 | Biuret–1.5% |
| Ammonium nitrate | $NH_4NO_3$ | 33.0 | 16.5 | 16.5 | - | 1.0 | - | 60 | - |

## APPENDIX XV
## NUTRIENT CONTENT, MOISTURE %, AND FREE ACIDITY OF PHOSPHATIC FERTILIZERS

| Fertilizer | Formula | $P_2O_5$ Content | | Moisture (%) | Free Acidity (%) | Remarks |
|---|---|---|---|---|---|---|
| | | Water soluble | Total | | | |
| Basic slag | $(CaO)_2 . (P_2O_5SiO_2)$ | - | 15.0–18.0 | - | - | |
| Rock phosphate | $[Ca_3(PO_4)_2]_3 . CaX_2$ | - | 20.0–38.0 | - | - | 95% passing in 0.15mm sieve |
| Single super phosphate | $Ca(H_2PO_4)_2$ | 16.0 | 18.0–20.0 | 12.0 | 4% as $P_2O_5$ | Sulfur–12% |
| Single super phosphate | $Ca(H_2PO_4)_2$ | 16.0 | 18.0–20.0 | 5.0 | 4% as $P_2O_5$ | Granulated |
| Triple super phosphate | $Ca(H_2PO_4)_2$ | 42.5 | 46.0 | 12.0 | 3% as $P_2O_5$ | Sulfur–1.5% |
| Bone meal | $Ca_3(PO_4)_2 (H_2O)_2$ | 8.0 | 20.0 | 8.0 | - | Raw |
| Bone meal | $Ca_3(PO_4)_2 (H_2O)_2$ | 16.0 | 22.0 | 7.0 | - | Steamed |

## APPENDIX XVI
## NUTRIENT CONTENT AND MOISTURE % OF POTASSIC FERTILIZERS

| Fertilizer | Formula | Water soluble | Moisture (%) | Remarks |
|---|---|---|---|---|
| Muriate of potash | KCl | 60.0 | 0.5 | Sodium as NaCl–3.5% |
| Potassium schoenite | $K_2SO_4.MgSO_4.6H_2O$ | 23.0 | 1.5 | Sodium as NaCl–1.5% |
| Sulfate of potash | $K_2SO_4$ | 48.0 | 1.5 | Total chloride as Cl–2.5%<br>Na as NaCl–2.0% |

## APPENDIX XVII
## ACID EQUIVALENT OF ACID FORMING FERTILIZERS

| Name of the fertilizers | Acid equivalent |
|---|---|
| Ammonium chloride | 128 |
| Ammonium phosphate | 86 |
| Ammonium sulfate | 110 |
| Ammonium sulfate nitrate | 93 |
| Ammonium nitrate | 60 |
| Urea | 80 |

## APPENDIX XVIII
## EQUIVALENT BASICITY OF BASIC FERTILIZERS

| Name of the fertilizers | Equivalent basicity |
|---|---|
| Calcium cyanamide | 63 |
| Calcium nitrate | 21 |
| Dicalcium phosphate | 25 |
| Sodium nitrate | 29 |

## APPENDIX XIX
## CA, MG, AND S CONTENTS OF SOME FERTILIZER MATERIALS

| | Nutrient content (%) | | | |
|---|---|---|---|---|
| Name of the fertilizers | Ca | Mg | S | Others |
| Ammonium sulfate | - | - | 24.2 | 21.0 (N) |
| Ammonium sulfate nitrate | - | - | 12.1 | 26.0 (N) |
| Basic slag | 33.9 | - | 3.0 | 15.6 ($P_2O_5$) |
| Calcium nitrate | 19.4 | - | - | - |
| Copper sulfate | - | - | 11.4 | 21.0 (Cu) |
| Elemental sulfur | - | - | 100.0 | - |
| Ferrous sulfate | - | - | 18.8 | 32.8 (Fe) |
| Ferrous ammonium sulfate | - | - | 16.0 | 6.0 (N)<br>16.0 (Fe) |

| Name of the fertilizers | Nutrient content (%) | | | |
|---|---|---|---|---|
| | Ca | Mg | S | Others |
| Gypsum | 40. | - | 23.5 | - |
| Potassium sulfate | - | - | 17.0–18.0 | 48.0 ($K_2O$) |
| Potassium chloride | - | - | - | 60.0 ($K_2O$) |
| Potassium magnesium sulfate | - | 11.1 | 22.3 | 31.0 ($K_2O$) |
| Rock phosphate | 39.0–48.0 | - | - | 25.0 ($P_2O_5$) |
| Super phosphate (single) | 25.0–30.0 | - | 12.0 | 16.0 ($P_2O_5$) |
| Super phosphate (triple) | 14.3 | - | 1.0 | 43.5 ($P_2O_5$) |
| Urea gypsum | 4.6 | - | 0.6 | 36.8 (N) |
| Urea sulfur | - | - | 10.0 | 40.0 (N) |
| Zinc sulfate | - | - | 17.8 | 36.4 (Zn) |

## APPENDIX XX
## SECONDARY AND MICRONUTRIENTS CONTENT OF FERTILIZER MATERIALS

| Nutrient | Name of the fertilizers | Nutrient content (%) |
|---|---|---|
| Calcium | Agricultural limestone | 80.0–95.0 ($CaCO_3$) |
| | Basic slag | 33.9 (CaO) |
| | Dolomite | 20.0–45.0 (CaO) |
| | Gypsum | 40.0 (CaO) |
| | Calcium ammonium nitrate | 10.0–20.0 (CaO) |
| | Super phosphate (single) | 25.0–30.0 (CaO) |
| | Rock phosphate | 39.0–48.0 (CaO) |
| Magnesium | Magnesium sulfate | 16.0 (MgO) |
| | Dolomite | 5.0–20.0 (MgO) |
| | Magnesite | 40.0 (MgO) |
| | Chelated magnesium | 2.0–10.0 (MgO) |
| Manganese | Manganese sulfate | 30.5 (Mn) |
| | Chelated manganese | 5.0–12.0 (Mn) |
| Sulfur | Ammonium sulfate | 24.0 (S) |
| | Super phosphate (single) | 12.0 (S) |
| | Potassium sulfate | 17.0–18.0 (S) |
| | Gypsum | 13.0–18.0 (S) |
| | Magnesium sulfate | 13.0 (S) |
| | Iron pyrites | 22.0–24.0 (S) |
| | Ammonium phosphate sulfate | 15.0 (S) |
| Molybdenum | Ammonium molybdate | 54.0 (Mo) |
| Iron | Ferrous sulfate | 20.0 (Fe) |
| | Chelated iron (FCO) | 12.0 (Fe) |
| | Chelated products | 5.0–10.0 (Fe) |
| Zinc | Zinc sulfate | 21.0 (Zn) |
| | Chelated zinc (FCO) | 12.0 (Zn) |
| | Zinc sulfate monohydrate | 33.0 (Zn) |

| Nutrient | Name of the fertilizers | Nutrient content (%) |
|---|---|---|
| Boron | Borax (deca) | 11.3 (B) |
| | Borax (penta) | 15.0 (B) |
| | Boronated SSP | 0.18 (B) |
| | Boric acid | 17.5 (B) |
| | Solubor | 20.5 (B) |
| Copper | Copper sulfate | 24.0 (Cu) |
| | Chelated copper | 5.0–12.0 (Cu) |
| Chlorine | Potassium chloride | 48.0 (Cl) |

## APPENDIX XXI
## GENERAL RECOMMENDED DOSES OF MICRONUTRIENT FERTILIZERS

| | Material and doses of application | |
|---|---|---|
| Micronutrient | Soil application | Foliar spray |
| Zinc | Zinc sulfate 25kg/ha | 0.5% zinc sulfate + 0.25% lime |
| Iron | Ferrous sulfate 50kg/ha | 1.0% ferrous sulfate + 0.50% lime |
| Copper | Copper sulfate 10kg/ha | 0.1% copper sulfate + 0.05% lime |
| Manganese | Manganese sulfate 10kg/ha | .0% manganese sulfate + 0.25% lime |
| Boron | Borox 10kg/ha | 0.2% borax |

## APPENDIX XXII
## SOME INDICATOR PLANTS OF NUTRIENT DEFICIENCY

| Nutrient | Indicator plants |
|---|---|
| N | Maize, cereal (small grain), mustard, apple, citrus |
| P | Maize, barley, lettuce, tomato |
| K | Potato, lucerne, beans, tobacco, cucurbits, cotton, tomato, maize |
| Ca | Lucerne, other legume crops |
| Mg | Potato, cauliflower |
| S | Lucerne, raya |
| Fe | Sorghum barley, citrus, peach |
| Zn | Maize, onion, citrus, peach |
| Mn | Apple, cherry, citrus, maize, oats, pea, radish, wheat |
| Cu | Apple, citrus, barley, maize, lettuce, oats, onion, tobacco, tomato |
| B | Lucerne, turnip, cauliflower, apple, peach |
| Cl | Lettuce |
| Mo | Cauliflower, other brassica sp, citrus, legumes, oats, spinach |

# Index

*Note*: Page numbers in *italics* refer to figures.